Common Sense Guide to Health and Safety for the Medical Professional

Subash Ludhra

Routledge
Taylor & Francis Group

LONDON AND NEW YORK

First published 2015
by Routledge
2 Park Square, Milton Park, Abingdon, Oxon OX14 4RN

and by Routledge
711 Third Avenue, New York, NY 10017

*Routledge is an imprint of the Taylor & Francis Group,
an informa business*

British Library Cataloguing-in-Publication Data
A catalogue record for this book is available from the British Library

Library of Congress Cataloging in Publication Data
Ludhra, Subash, author.
  Common sense guide to health and safety for the medical
  professional / Subash Ludhra.
    p. ; cm. — (Common sense guides to health and safety)
  I. Title. II. Series: Common sense guides to health and safety.
  [DNLM: 1. Health Personnel—standards—Great Britain.
  2. Occupational Health—legislation & jurisprudence—
  Great Britain. 3. Accidents, Occupational—prevention &
  control—Great Britain. 4. Infection Control—legislation &
  jurisprudence—Great Britain. 5. Safety Management—
  legislation & jurisprudence—Great Britain. WA 487.5.H4]
  RC965.M39
  363.11'961—dc23   2014020058

ISBN: 978-0-415-83546-6 (pbk)
ISBN: 978-1-315-85873-9 (ebk)

Typeset in Sabon
by Keystroke, Station Road, Codsall, Wolverhampton

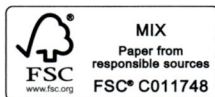

MIX
Paper from
responsible sources
FSC
www.fsc.org   FSC® C011748

Common Sense Guide to Health and Safety for the Medical Professional

An essential and short guide for employees who need to know more about health and safety within the medical profession. Whether it's for use alongside a training course or simply to brush up on your knowledge, it's perfect for equipping you with the principles of health and safety.

Friendly and accessible, this *Common Sense Guide* covers all the main aspects of medical health and safety in manageable chapters to provide you with the knowledge and understanding you need to look after yourself and others in the medical profession.

- Suitable for the non-health and safety professional
- Includes questions at the end of each module to consolidate your health and safety knowledge
- Certificate offered to those who complete the exam at the end of the book and return to be marked externally.

Subash Ludhra is a past president of The Chartered Institution of Occupational Safety and Health (IOSH) and considered to be an expert in the field of Risk Management. Having qualified as an Occupational Hygienist, Subash Ludhra now manages Arntara Management Ltd; an international risk management and loss control consultancy business that operates in the UK and overseas.

COMMON SENSE GUIDES TO HEALTH AND SAFETY

Common Sense Guide to Health and Safety at Work
978-0-415-83544-2

Common Sense Guide to Fire Safety and Management
978-0-415-83542-8

Common Sense Guide to Environmental Management
978-0-415-83541-1

Common Sense Guide to International Health and Safety
978-0-415-83540-4

Common Sense Guide to Health and Safety in Construction
978-0-415-83545-9

Common Sense Guide to Health and Safety for the Medical Professional
978-0-415-83546-6

Contents

FOREWORD vii

Welcome 1
How this guide works..7
How to complete the guide 11

MODULE ONE: Health and safety law and enforcement.................17

What is health and safety? 19
Why address health and safety? 21
The Health and Safety at Work Act 1974...................23
Employers' duties 24
Employees' duties...25
Duties of the self-employed...26
Suppliers/manufacturers...26
Health and safety law 26
Law enforcement...26
Penalties 31
Health and safety roles and responsibilities 32

MODULE TWO: Health and safety regulations 35

Safety signs 37
Electricity 39
Substances hazardous to health 41
First aid..45
Manual handling and people handling 46
Personal protective equipment (PPE)...............................51
Your workplace..54
Work equipment..54
Display screen equipment 60
Useful guidelines on how to use your workstation 63
Radiation...65

Contents

MODULE THREE: Commonly acquired infections and common hazards...71

About infection control...71
Methicillin-resistant *Staphylococcus aureus* (MRSA) 74
Clostridium difficille (C. diff)..75
Norovirus 76
Tuberculosis...77
Legionella...79
Gastroenteritis 81
The use of catheters 82
Enteral feeding 85
Falls from windows 87
Sharps 89
Clinical waste 91

MODULE FOUR: Accidents, incidents and infections 95

What is an accident/incident? 97
Why investigate accidents/incidents?.......................................101
Who suffers from an accident?...102
Hazards in the workplace 103
Occupational health..108

MODULE FIVE: Proactive health and safety 111

Health and safety policy and clear responsibilities 113
Audits 114
Inspections...114
Hazard/Near miss reporting 115
Training..115
Safe systems of work/Risk assessment..116
Plant/Equipment testing and inspection 118
Communication 118
Planned maintenance...118
Co-operation..119
Personal hygiene 119
Personal protective equipment...121

ANSWERS TO QUESTIONS 125
GLOSSARY/DEFINITIONS..133
IMAGE CREDITS 137
EXAMINATION 141

Foreword

The medical and healthcare industry presents unique health and safety challenges for all personnel involved in it. This can include exposure to various contamination-type risks as well as psychological and physical issues that may present risk of harm.

Workers in the medical and healthcare profession could be prone to injury and ill health through a host of factors which are unique to them because of being directly involved in the care of people.

As the Chief Executive Officer of a large farming and food production group, I am only too aware of the vital importance of getting health and safety right in an ever changing, dynamic working environment presenting an extensive range of risk factors.

We understand that, in order to control these risks, effective implementation of health and safety, in a systematic and targeted way, driven forward with enthusiasm and energy, can go a very long way to reducing risk of injury and ill health, no matter what that industry happens to be.

It is a proven fact in our business that a safe, well-planned and organised organisation is also a much more efficient one, viable and sustainable for the long-term future. I am sure that these considerations apply equally well across other sectors.

The medical and healthcare profession should need no convincing of the compelling case for good health and safety, and a significantly important part of that is the need for ongoing, high-quality information, instruction and training and ongoing learning and development of the people involved in that profession.

This new book is an excellent guide for those non-health and safety professionals who need to know more about health and safety in the medical and healthcare profession. This guide provides simple guidance and will support basic training without being overly theoretical or specific. It offers an easy reference for learning the basics of health and safety in the medical and healthcare sector.

It therefore provides a valuable new addition to our options to improve and develop our collective understanding of health and safety, and it follows that any improved learning gives us all an even greater chance to take the actions needed today and tomorrow to avoid the devastation that personal injury and ill-health causes.

A healthy and safe medical and healthcare profession will, of course, provide benefit for the people they serve and the standard of care that they can provide.

John Shropshire – CEO
The Shropshire Group

Welcome

It is often said that we all know what is right and what is wrong, what we should and should not do. Surely it's just "common sense". Unfortunately common sense is not as common as we would like to believe. The definition of common sense is "the ability to behave in a sensible way and make practical decisions". Most individuals and employers will at some time do things that are not sensible or practical and some will do this regularly.

This guide has been developed to help improve your knowledge of health and safety and infection control in the healthcare sector in a light-hearted way. It is designed to further heighten the common-sense element of your existing knowledge.

By taking the time to improve your knowledge and learn more about health and safety in the workplace you can:

- avoid having an accident
- help prevent accidents occurring to others
- make your workplace safer
- avoid contracting an infection yourself
- avoid passing on an infection to others
- help your employer maintain a clean place of work
- uphold the public's and your clients' expectations on workplace hygiene.

Every year, hundreds of people are killed while in hospitals or care premises and thousands more suffer from illnesses as a result of infections contracted within those premises. In addition to this, there is the unnecessary loss of life and the cost to the service providers is enormous and the damage to public confidence is staggering.

Most importantly most of these injuries, illnesses and infections can be avoided or minimised through simple hygiene practices.

DID YOU KNOW?

- In 2012 there were 1,646 deaths involving *Clostridium difficile* (*C. diff*) infection in England and Wales, 407 fewer than in 2011 (2,053 deaths) (Source: www.ons.gov.uk/ons/rel/subnational-health2/deaths-involving clostridium-difficile/2012/stb-deaths-involving-clostridium-difficile-2012.html. 12.03.14).
- The number of death certificates mentioning Methicillin-resistant *Staphylococcus aureus* (MRSA) fell by 20% from 364 in 2011 to 292 in 2012 (Source: www.ons.gov.uk/ons/rel/subnational-health2/deaths-involving-mrsa/2008-to-2012/stb—-mrsa.html 12.03.14).
- It is estimated that the NHS in England could save £150m and many hundreds of lives by tightening hygiene rules in hospitals and investing in infection control (Source: according to the spending watchdog the National Audit Office).
- At least 100,000 cases of hospital-acquired infections occur each year in England, with an estimated 5,000 deaths (Source: according to the spending watchdog the National Audit Office).
- This costs the NHS in the region of £1bn annually (Source: according to the spending watchdog the National Audit Office).

By working through this light-hearted guide you will learn more about health and safety in the workplace and as a result you will be better placed to recognise the hazards and dangers in your workplace or even your own home 24 hours a day.

Injured man

How this guide works

AIMS

The aim of this guide is to provide you with a basic under-standing of:

- the need to manage health and safety
- health and safety law and enforcement
- health and safety legislation
- accidents/incidents and what causes them
- hazards in the workplace
- commonly transmitted infections
- good personal hygiene practices
- general cleaning and hygiene
- proactive measures that can be taken to help reduce accidents and the spread of infections to ensure that you can work in a manner that is safe for you and your colleagues at all times.

OBJECTIVES

By the time you finish this guide, you will be able to:

- identify common hazards within your place of work
- define hazard and risk
- help your employer improve health and safety stan-dards at your place of work
- assess risks
- understand civil and criminal law relating to health and safety
- lift and move loads more safely
- be able to recognise and participate in a number of your employer's proactive safety measures
- recognise safety signs and understand their meaning
- reduce the likelihood of spreading infections

- wash your hands correctly
- work safely with electricity
- know what to do when handling sharps
- know what a reportable accident is and who to report it to
- have a greater understanding of health and safety and its importance in your everyday work and home life.

How to complete the guide

Before you start to complete this guide please read the guidance notes below in order to ensure that you get the most out of your training.

WHAT DO YOU NEED TO COMPLETE THE GUIDE?

You will need:

- a quiet, cosy environment that allows you to relax and make notes
- a pen and paper to make notes
- a desire to learn and improve your knowledge of health and safety.

GUIDANCE ON LEARNING

This guide has been produced to help you learn more about health and safety, reduce the risk of spreading infections in your workplace and to reduce the likelihood of you or your colleagues having an accident or suffering from ill health. The guide allows you to complete your studies at your own pace with the support of your manager or supervisor. However, we recommend that you complete the guide within four weeks.

Throughout the guide there are simple questions designed to help you test your subject knowledge and learn.

If you cannot answer the questions, please read the relevant topic again to refresh your memory. If you are still in doubt, please speak to your line manager who will be able to assist you.

When you have completed the guide and the exercises, you may wish to complete the examination (30-question multiple-choice exam paper) on pages 141–150. On achieving the required pass mark (75%), you will receive a certificate to confirm that you have completed the guide and passed the associated examination.

HOW TO USE THE GUIDE

The guide is divided into modules. We recommend that you complete one module at a time in full, starting with module one, progressing sequentially through to the last module. You do not have to complete the guide in one sitting. A lot of information is provided and you may learn more effectively by tackling the modules in bite-sized chunks.

The modules are designed to take you through a specific learning pattern to help you learn. Each module contains questions to make you think about your own job and workplace. At the end of each module there are a series of questions relating to the module you will have read through to test your knowledge and understanding. The answers can be found on pages 125–131 of the guide. There may be times when you feel you need help and support in completing the guide. Should this be the case please speak to your line manager.

WHEN DO I GET MY CERTIFICATE?

Once you have completed the guide and the examination paper, the paper will be marked and on achieving the required pass mark a certificate pdf will be emailed to you as soon as possible.

Remember the certificate only confirms that you have completed the guide and passed the associated exam. The real benefit to you will come from your improved knowledge and ability to identify hazards and reduce the risk of having an accident at work or in the home. **Good Luck!**

W' nning trophy

Health and safety law and enforcement

HOSPITAL TRUST SENTENCED OVER FATAL LEGIONNAIRES' DISEASE. . . AT LEAST SEVEN PATIENTS WERE INFECTED FROM THE HOSPITAL'S WATER SYSTEM OF WHICH TWO DIED

Welcome to module one. In this module you will learn about the importance of health and safety and the Health and Safety at Work Act.

WHAT IS HEALTH AND SAFETY?

Health and safety is about the measures necessary to control and reduce risks to an acceptable level, to ensure the health and/or safety of anyone who may be affected by the activities of people at work or others affected by your undertaking.

> "IN REALITY A WORKPLACE CAN NEVER BE 100% RISK FREE"

However, adequate controls and systems need to be in place that are acceptable to the employer, employees, clients and the enforcement authorities.

Any instruction given to you concerning health and safety is for your own well-being and must be adhered to at all times.

Remember successful health and safety management relies on the commitment of all parties involved (employers, employees, agencies and clients) and the systems can only be successful when all parties co-operate.

Employers and
employees must work
together

Health and safety management revolves primarily around
the term REASONABLY PRACTICABLE (i.e. as an
employer, did the company do what was reasonably practicable to safeguard their employees or others affected by
them?).

This is a balancing act of RISK (the probability of an event
occurring and the likely consequences if it does occur)
against COST (this may be in terms of money, time, effort,
latest technology and so on).

Risk   **Cost**

Reasonably practicable

If the RISK relates to slipping in the kitchen and the COST of removing/replacing the slippery surface is very high – then it may be acceptable to minimise the risks in other ways – through warning signs, coatings, the provision of appropriate safety footwear and so on. Generally the greater the risk the less important the cost element becomes.

WHY ADDRESS HEALTH AND SAFETY?

1. **To comply with statutory legislation**

You and your employer are required to comply with the requirements of the Acts and Regulations relevant to your employer's business (this will be covered in more detail later); examples of Acts and Regulations include:

- The Health and Safety at Work Act
- The Control of Substances Hazardous to Health Regulations
- The Food Safety Act
- The Food Safety (General Food Hygiene) Regulations
- The Management of Health and Safety At Work Regulations
- The Health and Safety (Sharps Instruments in Healthcare) Regulations
- The Health and Social Care Act
- The Personal Protective Equipment at Work Regulations
- The Manual Handling Operations Regulations
- The Health and Safety (Display Screen Equipment) Regulations
- The Workplace (Health, Safety and Welfare) Regulations
- The Provision and Use of Work Equipment Regulations
- The Health and Safety (Consultation with Employees) Regulations
- The Reporting of Injuries, Diseases and Dangerous Occurrences Regulations

- The Environmental Protection Act
- The Health and Safety (First Aid) Regulations.

This is only a small selection of the Acts and Regulations that organisations have to comply with.

2. To comply with internal company policies

Responsible employers have internal health and safety policies and procedures. These policies and procedures set out their own internal standards, which in some cases may exceed the requirements of the relevant regulations. You, as an employee, agree to comply with these as part of your terms and conditions of employment.

3. Moral obligations

To reduce the number of health and/or safety-related accidents and incidents and to safeguard you the employee and your patients/clients.

No employee or patient/client arrives at your workplace expecting to have an accident or contract an illness. Therefore by increasing your own awareness of health and safety you can help reduce the likelihood of accidents and incidents occurring and safeguard yours and your patients'/clients' well-being.

4. It's cost effective

All incidents cost money; for example, staff working overtime, the cost of employing agency staff, loss of production, failure to meet service requirements and extra pressure on the staff left to cover absenteeism. However, the real cost is the damage caused to the injured employee's, patient's or client's well-being and their suffering. This also affects their family and relations who are sometimes overlooked after a serious incident occurs.

Failing to manage your health and safety can affect your employee's, patients' or client's well-being

5. There is a business need to do so

To maintain your employer's good reputation, patients/ clients and the public expect your employer to have excellent health and safety and infection-control systems in place and to be able to demonstrate the promotion of a good, safe working environment.

> "A POOR SAFETY REPUTATION COULD COST YOUR EMPLOYER'S BUSINESS DEARLY"

THE HEALTH AND SAFETY AT WORK ACT 1974

The Health and Safety at Work Act was introduced in 1974 to help protect the large number of employees who, at the time, were not covered by any health and safety legislation. Its aims were:

- to secure the health, safety and welfare of persons at work;

- to protect other people from health and safety risks caused by the work activities – for example, patients, clients, contractors, other visitors on the premises or affected by you;
- to control the storage and use of explosive, flammable and dangerous substances (this has developed to become the Control of Substances Hazardous to Health Regulations);
- to control atmospheric emissions of certain substances that could prove to be harmful (which has now developed into the Environmental Protection Act).

More importantly it sets out clear definitions of responsibilities for employers, employees, the self-employed and suppliers/manufacturers.

EMPLOYERS' DUTIES

Employers have a duty under the Health and Safety at Work Act to:

- make provision and maintenance of plant and systems of work that are safe and without risk to health;
- ensure safety and absence of risks to health in connection with the use, handling, storage and transport of articles and substances;
- provide information, instruction, training and supervision to ensure the health and safety at work of employees;
- provide you with a place of work, which is in a condition that is safe with respect to access and egress;

Obey the law and your duties

- provide and maintain a working environment with adequate welfare facilities such as toilets/washing facilities;
- ensure that persons not in his employment, that is, patients, clients, visitors, contractors and so on, are not exposed to risks to their health and safety.

EMPLOYEES' DUTIES

As well as employers, you as an employee have a legal and moral obligation to:

- care for the health and safety of yourself and of other persons who may be affected by what you do or do not do;
- co-operate with your employer by complying with internal and external policies imposed for work activities to ensure your employer is able to fulfil his legal obligations.

You should not intentionally or recklessly interfere with or misuse anything provided in the interests of health, safety or welfare. Your employer may have provided safety equipment and procedures to help reduce risk in the workplace and your co-operation is required at all times.

A failure to meet your legal obligations could lead to internal disciplinary action and/or external enforcement action directly against any employee who is found to have broken a health and safety rule that results in injury, illness or property damage.

Tampering with or damaging safety equipment (e.g. machinery guarding, personal protective equipment, fire extinguishers) could lead to dismissal by your employer or prosecution from external enforcement authorities.

"REMEMBER: WORKING SAFELY MUST BE A CONDITION OF EMPLOYMENT"

DUTIES OF THE SELF-EMPLOYED

Under the Act the self-employed have a duty to ensure that their work activities do not endanger themselves or others.

SUPPLIERS/MANUFACTURERS

Suppliers and manufacturers of articles and substances have a duty to:

- ensure that the product designed or constructed is safe when properly used;
- test or have the product tested to ensure that it is safe;
- provide adequate information and instructions for the user.

HEALTH AND SAFETY LAW

There are two important areas of law in health and safety.

Civil Law Tends to deal with the area of compensation awarded or claimed when incidents occur or something goes wrong (e.g. clinical negligence).

Criminal Law Deals with the punishment awarded by the courts when the employer/employee breaks the law.

LAW ENFORCEMENT

The policing and monitoring of health and safety in the UK is generally carried out and enforced by Health and Safety Executive (HSE) Inspectors and Environmental Health Officers (EHOs). In addition, the Care Quality Commission

County court

Magistrates' court

(CQC) and other agencies have an important role to play (see below).

The HSE and EHOs have a wide range of powers, which include:

- visiting a place of work at any reasonable time
- carrying out investigations and examination

- dismantling and removing equipment
- taking samples of products
- taking photographs and viewing documents
- taking statements from key personnel
- requiring assistance.

Their primary function is to help to ensure employers are complying with the relevant regulations and safeguarding their employees and others affected by their business. However, while in their enforcement role they can also:

- Issue an **Improvement Notice**. This means something is not safe and they want it made safe; they will stipulate a reasonable time period to allow the employer to comply.
- Issue a **Prohibition Notice**. This means that something is so dangerous that the employer/employee must stop doing it immediately. The employer must then make it safe and be able to demonstrate to the enforcement body that it is safe before recommencing the activity.
- **Prosecute** employers and employees. This is generally only carried out when there is very strong evidence to suggest that an individual/company deliberately did something they should not have been doing.

The Care Quality Commission is the independent regulator of health and social care. This includes the care provided by the NHS, local authorities and voluntary organisations in registered settings. They regulate providers of healthcare services to people of all ages, including hospitals, ambulance services, clinics, community services, mental health services and other registered locations, including dental and GP practices, providers of social care services for adults in care homes (where nursing or personal care is provided), in the community and in people's own homes. They focus on people who find themselves in vulnerable circumstances, including those with mental health problems, learning disabilities, physical disabilities or long-term health conditions and older people and providers of services for people whose rights are restricted under the Mental Health Act. They inspect services and publish the results to help users make informed decisions about the care they receive.

The CQC generally use compliance actions in response to breaches of regulations with a minor impact on people, or where the impact is moderate but it's happened for the first time. Service providers are required to provide a report setting out how they intend to address the problem and the action they will take to become compliant. Compliance actions are often a precursor to enforcement action.

Enforcement action can be taken where the breach is more serious, or where a compliance action has not worked. Enforcement action can be under either **Civil Law** (to protect people from harm) or **Criminal Law** (to hold a registered provider or manager to account for causing harm or for breaking the law).

Under Civil Law, the CQC can issue a warning notice, impose or change a condition of registration, suspend registration or cancel registration.

Where people are at immediate risk of significant harm, the CQC can use their "urgent" powers, which allows them to take immediate action. They can also issue a warning or fixed penalty notice, offer a caution or prosecute an offender or a provider who is not registered when they should be.

In addition, the following agencies have an enforcement role in England, Scotland or Wales.

Monitor is an independent regulator for the health sector in England. Its role is to protect and promote the interests of people who use healthcare services. Its duties include, assessing whether NHS trusts are ready to become NHS foundation trusts and ensuring that foundation trusts are financially viable and well led, in terms of both quality and finances.

The Medicines and Healthcare Products Regulatory Agency (MHRA) is a government agency that is responsible for ensuring that medicines and medical devices work, and are acceptably safe.

Healthcare Improvement Scotland (HIS) incorporates the Healthcare Environment Inspectorate (HEI). Its focus is to reduce healthcare-associated infection risk to hospital patients, to improve the care of elderly patients and to regulate independent healthcare services through an inspection framework. HIS inspectors do this by carrying out announced and unannounced inspections in acute NHS hospitals in Scotland to check that the NHS QIS standards for older people in acute care and standards for prevention and control of healthcare associated infection are being met.

Social Care and Social Work Improvement Scotland (SCSWIS) scrutinises social care, social work and child protection services. Known as the "Care Inspectorate", it inspects, regulates and supports improvement of services and provides public assurance on service quality.

The Care and Social Services Inspectorate Wales (CSSIW) regulates social care, early years' services and local authority care support services.

Healthcare Inspectorate Wales (HIW) reviews and inspects NHS and independent healthcare organisations. Services are reviewed against a range of published standards, policies, guidance and regulations. It also registers independent services and has powers to take enforcement action in those matters. HIW is the Local Supervisory Authority (LSA) for the statutory supervision of midwives. It also has inspection and enforcement powers for the Ionising Radiation (Medical Exposure) Regulations.

The National Institute for Health and Care Excellence (NICE) provides guidance to support healthcare professionals and others to ensure that the care they provide is of the best possible quality and offers the best value for money. Their role is to improve outcomes for people using the NHS and other public health and social care services by:

- producing evidence-based guidance and advice for health, public health and social care practitioners;
- developing quality standards and performance metrics for those providing and commissioning health, public health and social care services;
- providing a range of information services for commissioners, practitioners and managers across the spectrum of health and social care.

There are several professional regulatory bodies who aim to ensure that proper standards are maintained by health and social care professionals and act when they are not. In order to practise in the UK, professionals are required to register with their appropriate body. These bodies fulfil similar functions for different professions across the UK. Their main duties are to maintain an up-to-date register of professionals; set and maintain standards for education, training and conduct; and investigate when these standards are not met or when a professional's fitness to practise is in doubt (examples include the GMC, NMC, GDC).

PENALTIES

In a magistrates' court the maximum fine for a health and safety offence (under the HSWA) is **£20,000** (per offence) and/or a **six-month prison sentence.**

In a crown court the maximum fine is **unlimited** and in addition a **two-year** prison sentence can be imposed.

Note: The Health and Safety Executive's inspectors can also charge for their time under the fee for intervention (FFI) scheme where they identify material health and safety breaches.

HEALTH AND SAFETY ROLES AND RESPONSIBILITIES

As well as the duty imposed on employers and all employees under the Health and Safety at Work Act, within your organisation a number of individuals may have been given specific health and safety duties and responsibilities by virtue of their job functions. These individuals help to ensure that safety is taken seriously and is properly managed.

Q

Do you know who has specific health and safety responsibilities in your workplace? (Write your answer here, if the answer is no then find out.)

EXERCISE 1

Self-assessment questions

1. What is the maximum fine you can receive at a magistrates' court?

a) £10,000
b) £15,000
c) £20,000
d) £30,000

2. What are the two important types of law used by the CQC?

a) Civil & Matrimonial
b) Criminal & Peaceful
c) Civil & Criminal
d) Peaceful & Matrimonial

3. Can an enforcement officer take equipment away from your workplace?

a) Yes
b) No

4. List four reasons for employers to address health and safety:

a) ....

b) ....

c) ....

d) ....

5. Give two reasons why employees should take health and safety seriously:

a) ....

b) ....

6. Who enforces health and safety law in the UK?

a) Factory inspectors and firemen
b) Policemen and firemen
c) Environmental Health Officers and Factory Inspectors
d) Firemen and Environmental Health officers

7. Health and safety law places a duty on:

a) Employers only
b) Employers and employees
c) Employers, employees and the self-employed
d) Nobody

You will find the answers on page 127.

Health and safety regulations

DEATH SCANDAL HOSPITAL TO BE PROSECUTED

PATIENT DIED WHEN NHS TRUST STAFF FAILED TO REALISE THEY SUFFERED FROM DIABETES

Welcome to module two. In this module you will learn more about specific health and safety regulations that are likely to apply to your place of work.

SAFETY SIGNS

You will no doubt have seen signs on display in your own place of work and/or other places of work. Signs cost money and give an important message. They are not displayed to make the workplace look pretty.

COLOUR	MEANING	EXAMPLE
BLUE	Mandatory (must do)	Wash your hands
RED	Prohibition (must not do)	No Smoking/Fire signage
YELLOW	Warning of hazards	Caution wet floor
GREEN	A safe condition	Fire escape route/First-aid box

Signs are produced in four key colours to ease identification. Each colour gives a particular message, which should always be adhered to.

Always know where the signs are, what they mean and most importantly ensure that you comply with their requirements.

Q

Think about some of the signs in your place of work; can you describe three?

ELECTRICITY

Electricity is used in almost all workplaces and is often referred to as the invisible killer as it can kill without being seen.

Electricity can cause:

- electric shock
- electrical burns
- death
- fires.

The Electricity at Work Regulations were introduced to protect people from dangers associated with electricity and to protect equipment from excesses of electricity.

Your employer must ensure that the electrical equipment on site is:

- safe to use
- properly maintained/tested
- easily isolated from the mains supply
- only worked on by competent people
- protected from excesses of current
- protected from adverse weather conditions or other hazards.

If using electrical equipment you must:

- ensure that the equipment you use is properly maintained
- be competent to use the equipment
- visually inspect it for signs of damage before use
- not use this equipment near water (e.g. baths and sinks).

Remember it is the electrical current (measured in amps) that kills, not the voltage. A very small current passing through your body could kill you.

When using portable appliances plugged into the mains, always ensure that they are protected by using an in-line circuit-breaking device and never handle electrical tools with wet hands.

In any case of a suspected electric shock always:

- call for help;
- switch off the power supply if possible and safe to do so, then move the person away from the electrical source using a non-metallic object (do not touch them directly if there is a risk of electrocution to you);
- if the casualty is breathing, place them in the recovery position and call for medical attention;
- if the casualty is not breathing, call for help and attempt resuscitation;
- never touch the casualty with bare hands unless the power supply has been isolated;
- only give first aid if you have been trained to do so.

Do not overload sockets/extension leads

Q

Do you use any electrical equipment at work or in the home? What precautions do you take prior to and during use?

_____

_____

_____

_____

_____

_____

_____

_____

_____

_____

_____

SUBSTANCES HAZARDOUS TO HEALTH

Substances hazardous to health are any materials, mixtures or compounds used at work or arising from work activities, which have the potential to cause harm to people's health in the form in which they occur in the work activity.

Examples include:

- cleaning products used on the premises
- toners used in photocopiers within offices

Cleaning chemicals

- waste bodily fluids
- bleach used as cleaning agents
- detergents and fabric conditioners used within a laundry setting
- glues and solvents used in floor laying.

They can be:

- liquids, solids, dusts, powders, gases, or vapours.

They may typically be:

- toxic, corrosive, harmful or irritant.

They can cause damage by:

- coming into contact with the skin and eyes
- entering the body through cuts in the skin

- being breathed in
- entering the body through the mouth.

The Control of Substances Hazardous to Health Regulations (COSHH) require your employer to protect you from exposure to hazardous substances by:

- carrying out an adequate assessment of the hazards;
- preventing/controlling your exposure to hazardous substances.

When carrying out the assessment your employer must:

- identify hazardous substances in the workplace
- identify who is at risk
- obtain the relevant Material Safety Data Sheets (MSDSs)
- evaluate the risk.

Once the assessments are complete they must:

- eliminate the risk of exposure or introduce appropriate and effective control measures;
- maintain the measures put in place;
- monitor the effectiveness of the control measures;
- record the assessments;
- inform you of all risks and provide suitable instruction, training and supervision.

Control measures that can be applied to reduce risk include:

- elimination
- substitution
- preventing exposure
- limiting exposure time
- providing local or general ventilation

- improving housekeeping
- training
- providing health screening
- providing appropriate personal protective equipment (PPE).

When handling substances hazardous to health always:

- ensure you have been made aware of the hazards and risks;
- store them in correctly labelled containers;
- wear any protective equipment provided by your employer;
- treat them with respect.

Q

Which substances hazardous to health do you use at work or in the home?

What precautions do you take?

FIRST AID

No matter how safe your place of work is, there is always the potential for someone to have an accident or fall ill. This is why first aid forms a vital part of any employer's health and safety management system.

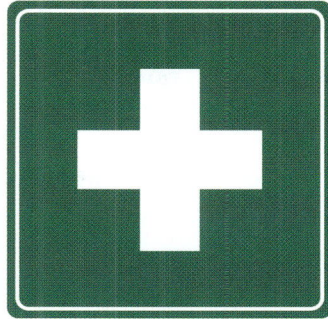

The Health and Safety (First Aid) Regulations were intended to ensure that in the event of an incident a qualified person would

Typical first-aid sign

be able to preserve life and prevent the deterioration of someone who has fallen ill or is the victim of an accident.

Your employer must assess the risks within your workplace and provide adequate cover.

In principle, your employer is required to have at least one fully qualified first aider per 50 employees (or emergency first aider as appropriate). That said, adequate cover must be provided during normal working hours taking into account shift patterns, holiday and sickness cover, the size of the location and hazards present. High-risk environments may require further specialist training for the first aider. In very low-risk environments an appointed person may be sufficient. It would be their responsibility to provide basic medical cover, call for medical assistance and maintain any medical supplies on site.

It is important that you ensure you know who your first aiders are, where their normal place of work is and how to contact them. In the event of an emergency, it could help save lives. If you are not sure ask your manager.

First-aid kits must contain as a minimum – individually wrapped plasters, sterile eye pads, triangular bandages,

sterile dressings and safety pins. First aiders are not authorised to issue or apply creams, pills, lotions, sprays or any
other medicines.

Q

Do you know who your local first aider is, where they are based and
how to contact them?

MANUAL HANDLING AND PEOPLE HANDLING

Manual handling is the lifting, lowering, pulling, pushing
or moving by hand or bodily force of an object or load. It
is a function you perform each and every day from brush-
ing your teeth to picking up a shopping bag.

Some job functions require more manual handling than others and it's not just the weight of a load that can cause problems. In fact the incorrect manual handling of any load can cause injury.

In the UK millions of working days are lost each year due to pains, strain and other injuries to the back. Manual handling injuries make up about a quarter of all injuries reported to the enforcing authorities.

Injuries can occur from a single event or may develop over a period of time.

The types of injuries suffered from manual handling are not confined to the back and can include cuts, hernias, wrenched shoulders, crushed feet or fingers, fractures and bruises.

The Manual Handling Operations Regulations were introduced to help reduce the number of accidents occurring within the workplace. In simple terms the regulations say that if manual handling can be avoided, then it should be. If it cannot be avoided, then your employer is required to carry out a risk assessment and introduce control measures to minimise the risk of injury to you.

The assessment must take into account:

- the **Task**, that is, what you are actually required to do;
- **individual capability**, that is, what you are capable of handling;
- the **Load**, that is, what it is you are required to handle;
- the **Environment**, that is, your surroundings.

Always remember the following procedure:

1. Examine the load – can the manual handling activity be avoided by redesigning or changing the task or avoided with the use of mechanical aids?

Ceiling hoist

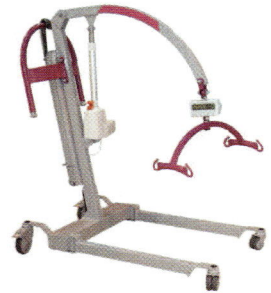
Mobile lifting hoist

2. Are you physically capable of handling the load or is assistance required?
3. Is protective clothing required to protect your hands, body or feet?

When lifting a load always:

1. Face the way you are walking or working – avoid twisting your trunk and overstretching at all times.
2. Position your feet so that they are approximately the width of your hips apart, with one foot slightly in front of the other and flat on the floor to provide a stable, balanced stance. Once the load has been lifted, the weight may be transferred to the front foot.

3. Always ensure you have a firm grip on the load, even if it means using gloves to prevent the load slipping.
4. Your back should be kept straight to maintain it in its natural and strongest position. To get down to the load, the knees and the ankles should be bent and the load raised gradually using the thigh and leg muscles.
5. The head should be kept up and the chin well in, as this helps to keep the spine in its natural position.
6. Arms should be kept as close to the body as possible, which helps to retain balance.
7. The body should be used to counterbalance the weight of the load.
8. If it is a team lift, then one person should control the lift so the lift is even and together.

When carrying a load always ensure:

1. You can see where you are going. Loads which extend to head height and which obstruct your vision are dangerous to you and to other persons nearby – if this is the case, you should either use a mechanical aid (sack barrow or pallet truck) or seek assistance from a colleague.
2. Your route is clear of any obstructions or slippery areas.
3. Particular care is taken when going around corners or negotiating stairs.
4. Sudden movements and twisting of the spine is avoided – face the way you are walking.
5. That if the load cannot be set down in the required position, it should be put down temporarily and re-lifted when the position is clear.
6. You do not try to change your grip while carrying – if your grip is slipping, put the load down and start again.

When putting down a load always ensure:

1. Your back is kept straight and in its natural position. If the load is to be lowered, the thigh and leg muscles must be used.
2. Your head is up and your chin tucked in.
3. You maintain a proper palm grip, but beware of trapping the fingers and hands beneath the load.
4. You use your body weight to counterbalance the load.
5. You keep your arms close to your body. Setting down and stacking should be only as high as it is possible to reach with the elbows tucked into the sides. Do not over-reach.
6. You maintain a firm, balanced stance, with your feet as close as possible to the centre of gravity of the load, and flat on the floor. If the load you are moving is too heavy or bulky, always seek assistance, which can be a colleague of similar height and build (one person should call the signals) or mechanical aids.

Sometimes it may be necessary to move a person or pick them up if they have fallen. Although this practice should be avoided, in the first instance, where available, mechanical aids should be used to assist you, but if these are not available, care must be taken to protect yourself and the person you are trying to handle – where possible seek assistance.

Your employer should have provided you with person-handling training for this type of activity.

All mechanical aids used to lift or move people should be regularly inspected (this may include a statutory inspection for some equipment) and the safe working load should not be exceeded.

PERSONAL PROTECTIVE EQUIPMENT (PPE)

Always wear the correct PPE

Personal protective equipment is any device or piece of equipment held or worn that can provide protection against one or more hazards.

When used correctly PPE can be used as an effective control measure against hazards. However:

"WEARING PERSONAL PROTECTIVE EQUIP-MENT DOES NOT REDUCE THE HAZARD AT ALL. THE HAZARD REMAINS THE SAME EVEN THOUGH THE RISK TO YOU IS REDUCED."

As such it must always be used as a last resort when all other control measures have been exhausted. The Personal Protective Equipment Regulations require your employer to carry out a risk assessment and identify when and where the wearing of PPE will be necessary.

When using PPE (which he must provide at no cost to you), your employer must take into account the following:

- its fit (i.e. is it the right size/shape for you?);
- its suitability against the risk;
- the ergonomics of the PPE (i.e. how well the PPE is designed for the individual user);
- any increase in risk to you or others as a result of using the PPE;
- its ongoing cleaning/maintenance;
- replacement procedures;
- storage facilities;
- adequate training to ensure you understand why it needs to be worn, the benefits it will provide to you, how it is to be maintained and stored and how you obtain replacements.

When provided, you as the employee are required to wear it and take reasonable care of it.

Questions to ask yourself when in your workplace:

- Is there a need for PPE for the job I am doing?
- Do I need protecting against any hazards?
- If I use PPE, is it suitable and does it serve its purpose?
- Is the PPE in good condition and cleaned regularly?
- Does it fit correctly and is it comfortable to work in?
- Is it readily available?
- Can I get replacements?
- Have I been trained on how to use the PPE? (On more complex equipment recorded evidence of training must be documented.)
- Does the PPE cause me any problems while I am working (heat rash or allergies)?
- Where do I store it when I have finished with it?

EXAMPLES OF PPE

Body part protected	Type of PPE available
Body	Overalls, coveralls, tabards
Feet	Shoes, boots
Hands	Gloves, gauntlets, wrist cuffs
Ears	Ear plugs, muffs
Eyes	Goggles, glasses
Face	Visor, hood, shield
Head	Head/hair cover
Skin	Barrier creams
Lungs	Breathing apparatus, respirators
Legs	Knee pads, leggings

Q

Do you or any of your colleagues wear PPE? If so, what is it worn for and how do you get it replaced?

YOUR WORKPLACE

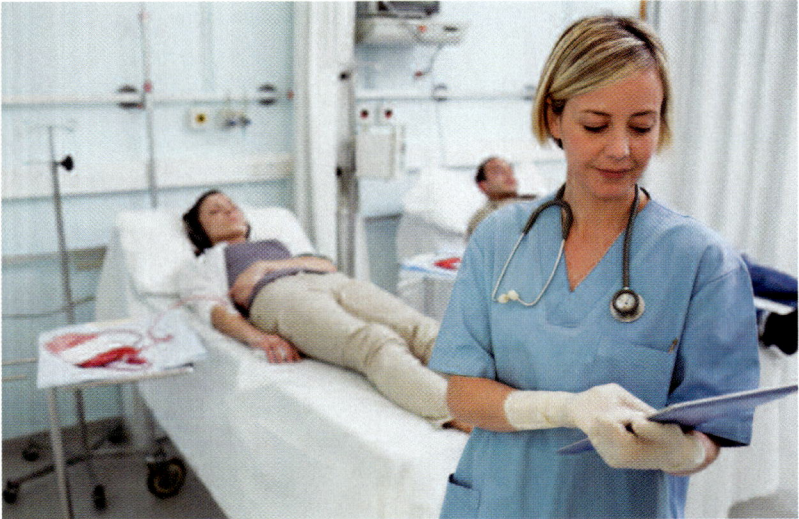

Hospital ward

Your employer is required to provide you with adequate facilities and ensure the workplace is safe and healthy.

Ventilation

Air must be provided either by natural (e.g. an open window) or mechanical means (e.g. a ducted fan-assisted system). In some cases the air has to be very clean for operating theatres and will have been cleaned using specialist filter devices.

Temperature

The workplace temperature must be at least 16°C (or at least 13°C where the work involves physical effort). If you work outside or in areas where it is unreasonable to maintain the minimum temperature or in areas that require temperatures that have to be kept high or low, then your employer may have to provide you with specialist clothing or make other suitable arrangements. It may be necessary to maintain a higher temperature due to the nature and needs of your clients.

Lighting

Your employer is required to ensure that there is suitable and adequate lighting for the tasks that you perform.

Stairways

All stairways should be kept clear and well lit. Guardrails must be fitted to open stairways and you should where possible ensure that you are holding a handrail when using the stairs.

Workstations

Your employer must ensure that you have adequate space to perform your job safely.

Furniture

Must be suitable for the tasks being performed.

Floors/Walkways

Must be kept clear and in good repair to prevent slips, trips or falls. Any spillage should be highlighted/cordoned off until it is cleared and the floor dried.

Housekeeping

The workplace must be kept clean and tidy. Equipment should be correctly stored and not left lying around.

Clear all spillages as quickly as possible and erect warning signs as necessary

Toilets

Adequate numbers of urinals and water closets must be provided. Where necessary additional toilets for disabled users should be provided and maintained. Where panic alarms are installed within toilets, they should be checked for operation on a periodic basis and the cord should be obvious and capable of reaching the floor.

Washbasins

The number required will be based on the number of male/female employees present. They must be suitably maintained with cold and warm water, soap and means to dry your hands. Temperatures should be regularly checked to ensure that users are not scalded.

Rest areas

May be required depending on the nature of the business. Where provided they should allow staff to rest comfortably.

WORK EQUIPMENT

Work equipment is defined as any machine or hand tool used at work. From this definition it is clear that almost all workplaces will have work equipment present. Although work equipment is generally there to make tasks easier, quicker or more efficient, if used incorrectly or without training, the work equipment can be dangerous and cause injury. Some types of work equipment (e.g. hoists and lifting equipment) may be subject to statutory inspection.

Under the Provision and Use of Work Equipment Regulations (PUWER) your employer is required to:

- provide suitable and safe work equipment;
- provide suitable and sufficient training on the correct use (including risks and precautions necessary) of the equipment for all authorised operators of the equipment.

There are a number of specific dangers associated with work equipment; these include:

Entanglement

Loose clothing or jewellery worn by you could become entangled within machinery.

Entrapment

Some machines can entrap parts of your body within their moving parts. Fingers, limbs and even the whole body can be pulled into the machine and crushed.

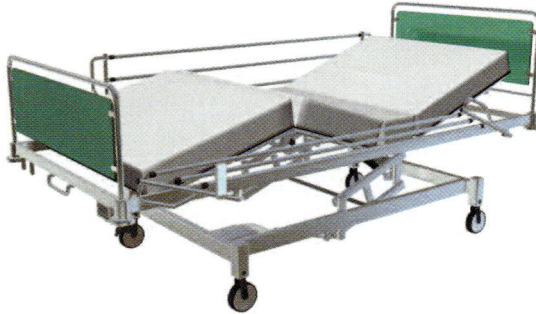

Bedside guarding

Contact

Moving parts of a machine could cause severe injuries to limbs or burns to the skin.

Ejection

Moving or rotating machines can throw out objects unexpectedly causing you injury.

Impact

You could be struck by a moving machine or an object being worked on by a machine.

Greater safety when using work equipment can be achieved by:

- buying and only using equipment that has been designed and constructed to remove foreseeable dangers;
- ensuring that equipment is used in a safe place not giving rise to danger;
- guarding dangerous parts of equipment to prevent injuries. Guards are designed and fitted to equipment to protect you and the equipment. Never tamper with guards, try to override them or operate a machine with the guards removed;
- never wearing loose clothing or jewellery around machinery;
- covering or tying long hair to prevent entanglement;
- ensuring adequate lighting so that everything can be clearly seen;
- ensuring that you wear any personal protective clothing deemed to be necessary;
- switching off and isolating the equipment when not in use and preventing unauthorised use;
- ensuring that the area around the work equipment is clean and tidy to prevent things or people falling;
- reporting all faults or suspected faults to your manager so that they can be investigated and where necessary rectified;
- only using work equipment if you are trained, competent and authorised to do so;
- never distract other people who are using work equipment.

When using hand-tools the following rules apply:

- Visually inspect the tools for signs of damage prior to use and report any defects.
- Always use the correct tool for the job. Do not attempt to make the tool fit the job.
- Ensure that the tools are in good condition.

- Use the tools in the correct way.
- Never use tools you have not been trained to use.
- Take out of service any broken or damaged tools identified.
- Work equipment can only ever be as safe as the operator using it.

DISPLAY SCREEN EQUIPMENT

The Display Screen Equipment Regulations were introduced to protect users of display screen equipment from the effects of:

- **Musculo-skeletal pains** – these are the aches and pains that can occur in the wrists, arms, neck, and upper/lower back as a result of poor posture or workstation design.

Display screen equipment workstation

- **Eye strain** – this can occur from poor screen adjustment, bad lighting or even from the use of incorrect glasses and is usually associated with headaches.
- **Mental fatigue** – this can occur as a result of poor workload planning. or limited understanding of the software or hardware being used. Your employer is required to assess all workstations (which includes the desk, chair, keyboard and working environment) to ensure that they comply with the requirements of the regulations and ensure the well-being of users of the equipment. If you are a user of display screen equipment, you may be asked to assist in the completion of the assessment form as part of the process.

Your workstation should meet the following criteria:

The screen

- Is the contrast and brightness adjustable?
- Are the characters of adequate size, stable and not flickering?
- Can the screen be easily swivelled and tilted to suit you?
- Is the screen free from glare or reflections?

The keyboard

- Is the keyboard detached from the bulk of the processor?
- Is the keyboard tiltable?
- Are all the keypad symbols legible?
- Is the surface of a matt finish?

Space

- Is there sufficient space to work comfortably?

Lighting

- Is there glare or a reflection on the screen from lighting or windows?
- Can the screen be moved to avoid the glare?
- Are there blinds to prevent reflections from sunlight?

Noise

- Is the workstation positioned alongside noisy machinery (e.g. printers)? If so can they be moved or covered?

Environment

- Is the workplace atmosphere too hot or cold as to make the user uncomfortable?

Cables

- Are all cables in a sound condition, not frayed or causing a trip hazard?

Seating

- Are there five castors on the chair?
- Can the back be adjusted to support the lower part of the back?
- Can both feet be placed flat on the floor? If not, is a footrest provided?
- Can the seat height be adjusted?

Work surface

- Is it of low reflection and large enough to accommodate any other piece of equipment needed (e.g. document holder, calculator, telephone)?

If your workstation does not comply with the above or should you have any concerns, please contact your manager for guidance.

USEFUL GUIDELINES ON HOW TO USE YOUR WORKSTATION

Seating

- Avoid slouching, sit upright and maintain the natural curve of your back.
- Adjust the backrest of your chair to support your lower back.
- Sit back in your chair to gain full support.
- Arrange your workstation to help maintain an upright posture.
- Don't sit at your workplace for long periods of time – try to plan your workload to have breaks away from the screen by doing other chores.
- Use a footrest if your feet cannot sit comfortably on the floor.

Upper body

- Adjust your seating so that your forearms are in a horizontal position.
- Align your hands with your forearms so you are working with straight supported wrists.
- Adjust the angle of your screen to suit your own sitting position and eye level.

Vision

- Position your screen so that reflections from windows (sunlight) or internal lighting are not causing problems on your screen. Window blinds may also be necessary.

- Position your screen so that the need to move your head and neck is minimised.
- If you use a document holder, position it alongside your screen to reduce movement of the eyes, head and neck.
- Adjust the screen to minimise the reflections from the lights, windows or other bright surfaces.
- Regularly clean your screen with the equipment provided.
- If you are identified as a user of DSE, then you may be entitled to an eye test paid for by the company.

In order to achieve the above it may be necessary to re-arrange furniture.

Remember the person using the equipment before you may have been of a different size and height.

Incorrect use of your workstation and equipment can lead to physical discomfort (e.g. backaches, wrist problems), eye strain (blurred vision, headaches) and stress.

If in doubt, speak to your supervisor or manager for assistance in making adjustments to your workstation.

Your PC may have connections to a network system. Do not unplug any cables unless you have checked with your system administrator.

Laptops

- If you use a laptop for work, then it should ideally be used with a docking station if being used for long periods of time.
- Do not use laptops on your lap.
- Try to use external mouse devices.
- Ensure it is used on a suitable desk so that you can maintain a reasonable body posture.

- Remember that carrying a laptop around constitutes manual handling so think about how it is being carried.

RADIATION

Radiation is a general term referring to any sort of energy that can travel through space either as a wave or a particle. Radiation may be classed as non-ionising radiation (low energy – e.g. ultraviolet radiation, visible light, infrared radiation, microwaves) and ionising radiation (high energy – e.g. X-rays).

Overall, there is little evidence to suggest most types of non-ionising radiation are harmful at levels people are normally exposed to, but some forms of non-ionising radiation are potentially dangerous.

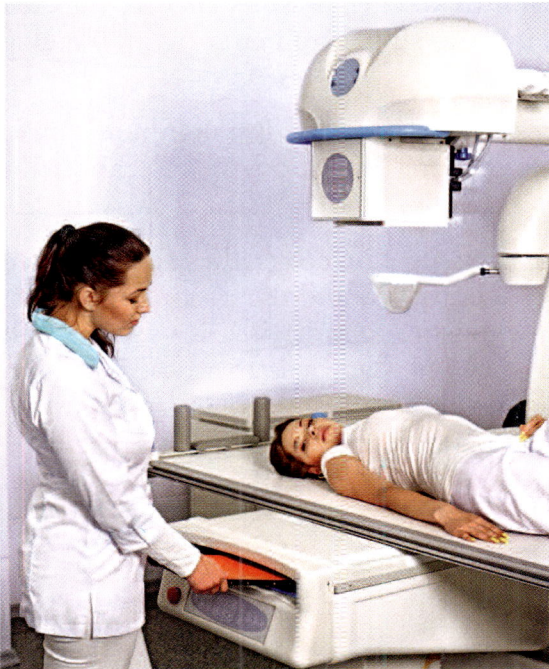

X-ray machine

The main proven danger of non-ionising radiation is damage to the skin caused by ultraviolet (UV) light. UV light primarily comes from the sun, but is also produced by sunbeds and sunlamps.

Ionising radiation is a more powerful form of radiation than non-ionising radiation and is therefore more likely to cause damage to cells. Exposure to ionising radiation can increase the risk of cancer and high doses can cause serious damage, including radiation burns. One of the most common sources of exposure to man-made ionising radiation is during medical tests or treatments such as X-rays and CT scans which use small amounts of ionising radiation.

Ionising radiation in the form of gamma rays is also common in what is known as nuclear medicine. This is a branch of medicine where radioactive substances can be used to help diagnose, and sometimes treat, a condition – for example, a mild radioactive substance is injected into the blood stream so it shows up better on an imaging scan.

In terms of treatments, radiotherapy is a common cancer treatment that uses ionising radiation to kill cancerous cells.

However, while it may sound dangerous, the radiation used in medicine is closely controlled and the risk of any problems resulting from exposure to radiation is very small. It is, however, important that if you use this type of equipment or are in an area where it is used, you must follow all safety instructions and adhere to rules in place.

Q

Do you use any equipment containing a radiation source? If so, what precautions are necessary?

EXERCISE 2

Self-assessment questions

1. What are the four colours used for safety signs?

a) ....

b) ....

c) ....

d) ....

2. Is it good practice to lift an adult person who you find fallen to the floor on your own?

a) Yes
b) No

3. What are the two types of radiation?

a) ....

b) ....

4. List four different health and safety control measures that can be applied to reduce risk.

a) ....

b) ....

c) ....

d) ....

5. Which muscles should be used when lifting a load?

    a) Arm muscles

    b) Leg muscles

    c) Back muscles

    d) Stomach muscles

6. What is the purpose of First Aid?

....

....

....

7. List two things that must be assessed when carrying out a VDU assessment

    a) ....

    b) ....

You will find the answers on pages 127–128.

Commonly acquired infections and common hazards

HSE WARNS HOSPITALS THAT THEY HAVE A DUTY TO ENSURE THE SAFETY OF PATIENTS FROM FALLS FROM WINDOWS FOLLOWING THE DEATH OF A PATIENT WHO FELL FROM A HOSPITAL WINDOW

Welcome to module three. In this module you learn more about commonly acquired infections and other hazards in hospitals or care-providing premises, what they are, what they do, how they occur and what needs to be done to prevent or control them.

ABOUT INFECTION CONTROL

Infections are caused by germs such as bacteria, fungi or viruses entering the body. They can be minor and stay in one area, such as a boil, or they can spread throughout the body, such as flu. Often, infections are easily dealt with, but sometimes they can cause serious problems. The recommendations on infection control focus on preventing infections that are associated with healthcare – for example, ways of preventing germs being spread on the hands of a healthcare worker or a carer.

The main forms of attack for hospital-acquired infections are:

- blood infections
- infections following surgery
- infections of the urinary system
- chest infections
- skin infections.

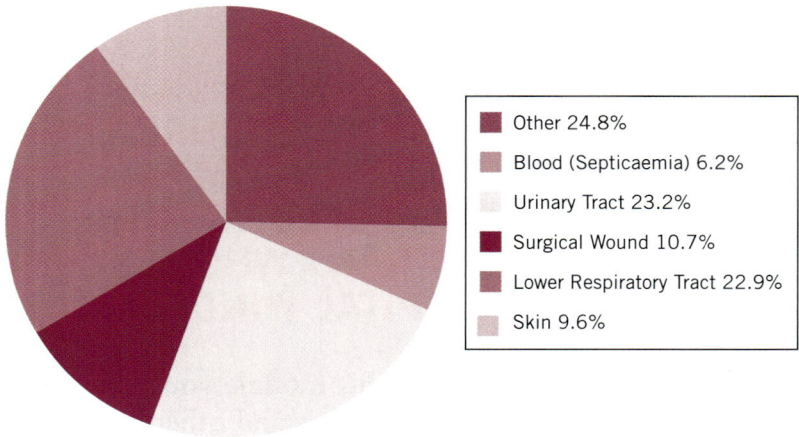

Other 24.8%

Blood (Septicaemia) 6.2%

Urinary Tract 23.2%

Surgical Wound 10.7%

Lower Respiratory Tract 22.9%

Skin 9.6%

METHICILLIN-RESISTANT *STAPHYLOCOCCUS AUREUS* (MRSA)

Staphylococcus aureus is a bacterium that is a common coloniser of human skin and mucosa. *Staphylococcus aureus* can cause disease, particularly if there is an opportunity for the bacteria to enter the body.

Illnesses such as skin and wound infections, urinary tract infections, pneumonia and bacteraemia (blood stream infection) may then develop. It can also cause food poisoning. Most strains of this bacterium are sensitive to many antibiotics, and infections can be effectively treated. Some *S. aureus* bacteria are resistant to the antibiotic methicillin and are termed methicillin-resistant *Staphylococcus aureus* (MRSA).

MRSA was relatively rare in the early 1990s representing only about 1–2% of serious infections caused by this species; however, this figure rose to more than 45% within 10 years.

MRSA is particularly feared in hospitals as it can cause an extremely wide range of serious disease such as pneumonia, septicaemia, bone infections and toxic-shock and it can only be reliably treated with vancomycin – an antibiotic that is usually given intravenously over a period of several days.

Vancomycin has several side-effects and is relatively toxic; it is also poorly absorbed by the body, causing problems in treating deep-seated MRSA infections and pneumonia.

CLOSTRIDIUM DIFFICILE (C. DIFF)

Clostridium difficile infection is the most important cause of hospital-acquired diarrhoea. *Clostridium difficile* is an anaerobic bacterium that is present in the gut of up to 3% of healthy adults and 66% of infants. However, *Clostridium difficile* rarely causes problems in children or healthy adults, as it is kept in check by the normal bacterial population of the intestine. When certain antibiotics disturb the balance of bacteria in the gut, *Clostridium difficile* can multiply rapidly and produce toxins which cause illness.

Clostridium difficile is a major cause of hospital-acquired diarrhoea in many countries where it can cause life-threatening illness, especially in the elderly and patients with severe underlying disease. *Clostridium difficile* was responsible for a fatal outbreak at Stoke Mandeville hospital among elderly patients. Comparisons between this organism and MRSA highlight the difficulties faced by healthcare workers in reducing the number and severity of hospital outbreaks.

Clostridium difficile is commonly found in the large intestine and infections usually occur following long-term antibiotic therapy that kills other bacterial competitors allowing *Clostridium difficile* to take over. It produces two major toxins that inflame the colon causing diarrhoea.

Contamination of the hospital environment from this source is key in causing and prolonging outbreaks as the bacteria produce spores that can survive on wet or dry surfaces in hospitals for long periods.

Clostridium difficile can be readily treated using vancomycin or metronidazole but importantly it is not killed by alcohol handwashes used by healthcare workers to prevent the spread of MRSA and other infections and is best dealt with by using soap and water.

NOROVIRUS

Norovirus, sometimes known as the 'winter vomiting bug', is the most common stomach bug in the UK, affecting people of all ages. It has an incubation period of between 24 and 48 hours. It is highly contagious and is transmitted by contact with contaminated surfaces, an infected person, or consumption of contaminated food or water. The symptoms of norovirus are very distinctive – people often report a sudden onset of nausea followed by projectile vomiting and watery diarrhoea.

An infection with norovirus is self-limiting and most people will make a full recovery in 1–2 days. It is important to keep hydrated – especially for children and the elderly.

Sufferers do not normally need to visit either a hospital A&E department or their GPs with symptoms as they risk spreading the virus.

The virus can spread person to person by the faecal–oral route, infection from aerosols of projectile vomit, environmental contamination (especially of toilets) and ingestion of contaminated food and water (especially bivalve molluscs).

It's not always possible to avoid getting norovirus, but ensuring good hygiene can help limit the spread of the virus. This can be achieved by:

- washing hands frequently and thoroughly with soap and water, particularly after using the toilet and before preparing food;
- avoiding the sharing of towels and flannels;
- disinfecting any surfaces or objects that could be contaminated with the virus;
- washing any items of clothing or bedding that could have become contaminated with the virus. Wash the items separately and on a hot wash to ensure that the virus is killed;
- flushing away any infected faeces or vomit in the toilet and cleaning the surrounding toilet area;
- avoiding eating raw, unwashed produce and only eat oysters from a reliable source (oysters have been known to carry the norovirus);
- not relying on alcohol gels as these do not kill the virus;
- using gloves when carrying out cleaning activities.

If you have norovirus, avoid direct contact with other people, and the preparation of food for others until at least 48 hours after your symptoms have disappeared. You may still be contagious even though you no longer have sickness or diarrhoea.

TUBERCULOSIS

Tuberculosis (TB) is a bacterial infection spread through inhaling tiny droplets from the coughs or sneezes of an infected person. Although a serious condition, it can be cured with proper detection and treatment.

TB that affects the lungs is the only form of the condition that is contagious and usually only spreads after pro-longed exposure to someone with the illness.

In most healthy people, the immune system (the body's natural defence against infection and illness) kills the bacteria. However, sometimes the immune system cannot kill the bacteria, but manages to prevent it from spreading in the body. The person will not have any symptoms, but the bacteria will remain in their body. This is known as latent TB.

If the immune system fails to kill or contain the infection, it can spread to the lungs or other parts of the body and symptoms will develop within a few weeks or months. This is known as active TB. Latent TB could develop into an active TB infection at a later date, particularly if the person's immune system becomes weakened.

TB is spread when a person with an active TB infection in their lungs coughs or sneezes and someone else inhales the expelled droplets containing TB bacteria. Although it is spread in a similar way to the cold or flu virus, TB is not as contagious. A person would usually have to spend prolonged periods in close contact with an infected person to catch the infection or have a weakened immune system because of HIV, diabetes or other medical conditions or be on long courses of medication, such as corticosteroids or chemotherapy.

Typical symptoms of TB include:

- having a persistent cough for more than three weeks that brings up phlegm, which may be bloody
- weight loss
- night sweats
- high temperature (fever)
- tiredness and fatigue
- loss of appetite.

In order to reduce the likelihood of spread of the virus, staff working in high-risk areas should be immunised and wear suitable masks.

Persons known or suspected to have TB should be provided with single-room accommodation, ideally with separate use of sanitary facilities. It may be appropriate to use negative air pressure rooms.

LEGIONELLA

Legionellosis is the collective name given to the pneumonia-like illness caused by legionella bacteria. This includes the most serious Legionnaires' disease, as well as the similar but less serious conditions of Pontiac fever and Lochgoil-head fever. Legionnaires' disease is a potentially fatal form of pneumonia and everyone is susceptible to infection. However, some people are at higher risk, including people over 45 years of age, smokers and heavy drinkers, people suffering from chronic respiratory or kidney disease and anyone with an impaired immune system.

The bacterium *Legionella pneumophila* and related bacteria are common in natural water sources such as rivers, lakes and reservoirs, but usually in low numbers. They may also be found in purpose-built water systems such as cooling towers, evaporative condensers and whirlpool spas. When conditions are favourable, the bacteria may grow, increasing the risks of Legionnaires' disease.

The early symptoms of Legionnaires' disease include a 'flu-like' illness with muscle aches, tiredness, headaches, dry cough and fever. Sometimes diarrhoea occurs and confusion may develop. Deaths occur in 10–15% of the general population and may be higher in some groups of patients. The incubation period can range from 2 to 19 days with a median of 6 to 7 days after exposure.

People become infected when they inhale legionella bacteria which have been released into the air in aerosolised form from a contaminated source. Once in the lungs the bacteria multiply and cause either pneumonia or a less serious flu-like illness (Pontiac fever). Certain conditions increase the risk from legionella, including:

- water temperature between 20 and 45°C, which is suitable for growth;
- creating and spreading breathable droplets of water (e.g. aerosol created by a cooling tower, or water outlets);
- stored and/or re-circulated water;
- a source of nutrients for the organism (e.g. presence of sludge, scale or fouling).

Control and prevention of the disease is through treatment of the source of the infection, that is, by treating the contaminated water systems, and good design and maintenance to prevent growth in the first place. It is therefore important to ensure that the water systems at your premises

Keep air conditioning towers maintained

have been properly risk assessed and a management system is in place. Once someone is infected, treatment is usually via a course of antibiotics.

Where thermostatic control valves are used to control water temperatures at tap and shower outlets, these should be regularly checked to ensure they are set at the correct temperature to minimise the risk of scalding users. If these valves are not in place, other control measures will be necessary to prevent the risk of scalding.

GASTROENTERITIS

Gastroenteritis is an infection of the stomach and bowel. The most common symptoms are vomiting and repeated episodes of diarrhoea. The two most common causes of gastroenteritis in adults are the norovirus and food poisoning.

Most types of gastroenteritis are highly infectious. The condition is mainly spread when bacteria found in faeces are transferred to the mouth.

Bacteria are commonly transferred through poor hygiene (if someone does not wash their hands after going to the toilet, any viruses or bacteria on their hands will be transferred to whatever they touch, such as a glass, kitchen utensil or food).

If you touch a contaminated object and then touch other things, you could be spreading the virus. If you touch your face, or if you eat contaminated food, you may become infected by the virus or bacteria.

As gastroenteritis is highly infectious, it is important to take steps to prevent it from spreading to other people. These include:

- washing your hands thoroughly after going to the toilet and before eating or preparing food;
- cleaning the toilet, including the handle and the seat, with disinfectant after each bout of vomiting or diarrhoea;
- not sharing towels, flannels, cutlery or utensils with others;
- not returning to work until 48 hours after your last bout of vomiting or diarrhoea.

THE USE OF CATHETERS

A catheter is a hollow tube that drains urine from a person's bladder into a special drainage bag (a catheter valve may be used instead of a bag; this is opened at regular intervals to drain the urine from the bladder).

An indwelling catheter is one that's in place all the time. An intermittent catheter is inserted at regular intervals or when the patient feels the need to urinate.

Usually, the catheter is inserted through the urethra (the tube where urine normally comes out). Sometimes a catheter is inserted into the bladder through a specially made hole in the side of the abdomen (this type of catheter is called a "suprapubic catheter"). A small balloon keeps the catheter in place inside the bladder.

It is possible to get an infection when using a urinary catheter. Bacteria get into the urethra from the drainage bag, or at the point where the tube enters the body.

Medical staff and carers should ensure that the correct procedures are followed when fitting catheters and that patients are correctly briefed in the ongoing use of them.

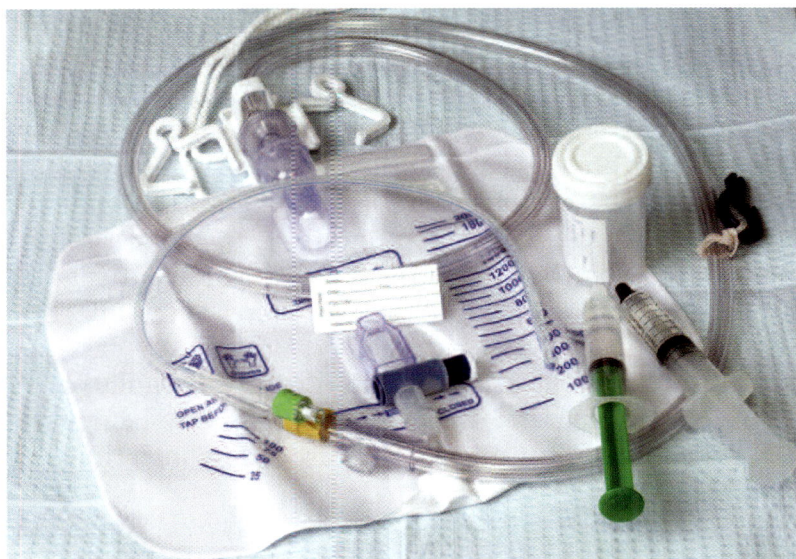

Follow the correct procedures when fitting or removing catheters

Good practice guidance includes the following:

- Before and after handling the catheter or drainage bag, the place where the catheter enters the body must be cleaned daily with soap and water and dried.
- The tubes of the catheter should not be allowed to become blocked.
- When inserting an intermittent catheter, the appropriate lubricant (some intermittent catheters are supplied ready lubricated) should be applied. The lubricant makes it easier to put the catheter in place and helps avoid infection. If lubricant sachets are being used, they should be used once then thrown away. Containers or tubes of lubricant can be used more than once, but should only be used by one patient.
- A reusable intermittent catheter should be washed after use with water, dried and stored according to the manufacturer's instructions.

- Ensure the drainage bag or catheter valve is connected to the catheter at all times, except when changing the bag. This "closed system" reduces the risk of infection.
- At night, ensure a special night drainage bag is added without breaking the closed system.
- Ensure the drainage bag is lower than the patient's bladder to allow urine to drain.
- Don't let the bag touch the floor when the patient is in bed or resting, because this can increase the infection risk.
- Ensure that the drainage bag is emptied regularly.
- Ensure the patient drinks plenty of fluids to help maintain a good flow and prevent ascending infections.

Central venous catheters

A central venous catheter (CVC) is a tube that is put into a major vein, normally in the chest or neck. (A vein is a blood vessel that carries blood to the heart.) There are many reasons why people may have a CVC. They may need

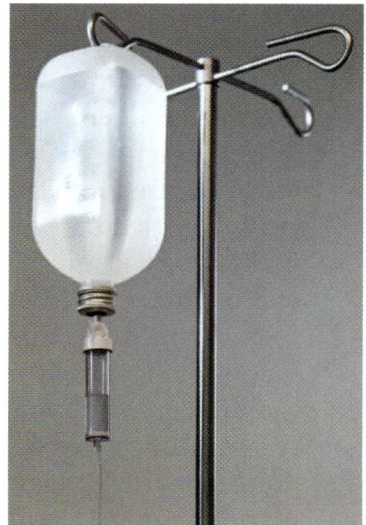

Follow the correct procedures when fitting or removing catheters

blood products, liquid drugs, food or other fluids delivered slowly into their bloodstream. Some people may need to use a CVC for a long time or for life.

Avoiding infection

Because a CVC is put into a major vein, serious infections can occur very quickly. People with CVCs, their carers and healthcare workers need to follow strict rules to prevent infection.

Important points about central venous catheters

- Ensure correct instructions are followed at all times.
- Ensure hands are washed carefully with soap and water or a handrub solution before touching the CVC.
- Ensure sterile gloves are worn when touching the insertion site or changing the dressing.
- Ensure that the dressing at the insertion site is changed every seven days or sooner if necessary (e.g. if it becomes wet, dirty or loose) using the correct cleaning solution and dressings.
- Do not put any cream, ointment or solution on the insertion site, unless it has been specifically prescribed for the patient.
- Ensure the catheter and its entry points are cleaned as instructed, before and after use with the correct solutions.

ENTERAL FEEDING

Enteral feeding, sometimes called enteral nutrition or artificial feeding, is required for adults and children who cannot eat normally. Liquid feed is given through a fine tube that enters the body by one of three ways:

- through the nose into the stomach – naso-gastric feeding
- directly into the stomach – gastrostomy or PEG feeding
- directly into the small bowel – jejunostomy feeding.

Some people who have serious problems with their digestive system may need to receive feed through a tube for a long time or even for life.

Medical staff and carers should ensure that the correct procedures are followed when fitting the tubes and that patients are correctly briefed in the ongoing use of them. Avoiding infection is very important for people who are on enteral feeding because infections such as gastroenteritis (stomach upset) can occur.

Good practice guidance

Careful and regular hand washing is very important. Whenever possible, pre-packaged feed that is ready to use and does not need mixing or diluting should be used. If feed has to be prepared, it is important not to touch it directly with your hands, and to use a clean working area and clean equipment when preparing it. Cooled boiled water or fresh sterile water (not bottled mineral or table water) should be used to mix the feed. It may be prepared up to 24 hours in advance and kept in the fridge (if the manufacturer's instructions say it is alright to do so).

Important points about enteral feeding

- Ensure that the feed is stored in accordance with the manufacturer's instructions.
- Ensure hands are washed thoroughly before preparing the feed or touching the equipment.

- Avoid any unnecessary handling of the equipment.
- Ensure that the place where the tube enters the body (insertion site or stoma) is cleaned with water every day and properly dried.
- The enteral feeding tube should be flushed with fresh tap water before and after feeding or administering medications to avoid blockages. Enteral feeding tubes for patients whose immune systems are not functioning properly (who are "immunosuppressed") should be flushed with either cooled freshly boiled water or sterile water (not bottled mineral or table water) from a freshly opened container.
- Avoid handling the feeding system more than necessary when connecting it to the tube.
- Ensure all waste is disposed of after each feeding session.
- Regularly inspect the insertion site for signs of infection.

FALLS FROM WINDOWS

From time to time persons fall out of windows within buildings and the risk increases with the number of floors within the building. Three broad categories of falls have been identified by the HSE:

1. Accidental falls.
2. Falls arising out of a confused mental state.
3. Deliberate self-harm.

Accidental falls are a minority, but can occur where a person is sitting on a window sill, or where the sill height is low and acts as a pivot, allowing them to fall out.

A significant number of reports refer to the mental state of the patient. Senility, dementia, mental handicap or illness, the effect of drink and drugs (both prescribed and illegal) can all cause anxiety and confusion. Individuals are inclined to try to escape from a perceived hostile environment, or to use a window believing it to be a door, possibly unaware that they are not at ground level. Other influencing factors include patients being unfamiliar with new surroundings, often exasperated by uncomfortable temperatures, anxiety, broken sleep and medication effects.

Deliberate self-harm is a recognised risk for patients with certain medical conditions.

Hospitals and care providers have a duty to ensure the safety of their patients, especially from the dangers of open windows. They need to ensure that vulnerable patients in their care are not put at risk. The risk of someone getting out of an unsecured window should be properly assessed and systems put into place to ensure that window restrictors are properly fitted and applied. In addition, the

Ensure patients cannot fall from windows

propensity to self-harm should be considered as part of the initial clinical assessment, particularly for psychiatric patients, and actioned as appropriate.

SHARPS

Sharps are anything that might cut, graze or prick you or others such as needles, lancets, scalpels, stitch cutters, glass ampoules, sharp instruments and broken crockery and glass.

Sharps must be handled and disposed of safely to reduce the risk of exposure to blood-borne viruses. Always take extreme care when using and disposing of sharps.

Whenever possible avoid using sharps. Where they cannot be avoided, then the following should be applied:

Clinical sharps should be single-use only.

Sharps should not be passed directly from hand-to-hand and handling should be kept to a minimum.

Needles should not be re-capped, bent, broken or disassembled before use of disposal.

Needle safety devices must be used where there are clear indications that they will provide safer systems of working for healthcare personnel.

Sharps containers must conform to agreed standards (UN3291 or BS7320).

Assemble sharps containers by following the manufacturer's instructions.

Label sharps containers with the source details.

Used sharps must be discarded into a sharps container at the point of use by the user.

Sharps containers must not be filled above the mark indicated on the container.

The aperture of the sharps container should be closed when carrying or, if left unsupervised, to prevent spillage or tampering.

Place sharps containers on a level, stable surface.

Carry sharps containers by the handle – do not hold them close to the body.

Never leave sharps lying around.

Do not try to retrieve items from a sharps container.

Do not try to press sharps down in the sharps box to make more room.

Lock the container when it is three-quarters full using the closure mechanism.

Place damaged sharps containers inside a larger container – lock and label prior to disposal. Do **not** place sharps inside a waste bag.

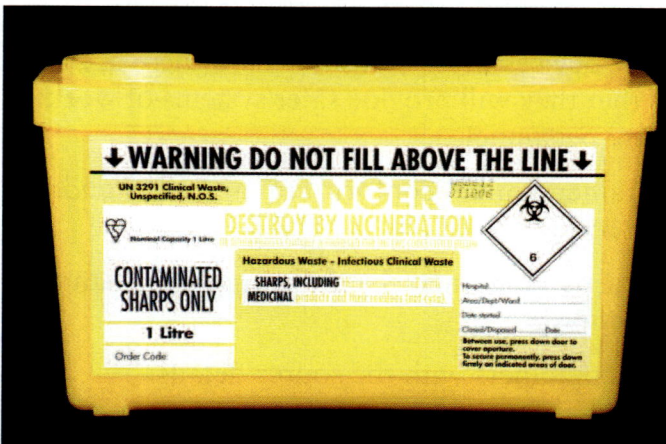

Sharps box

Containers in public areas must be located in a safe position, and must not be placed on the floor.

Always report any sharps injuries.

CLINICAL WASTE

Clinical waste consists of four main categories:

1. Human or animal tissue, blood or bodily fluids, or excretions.
2. Dressings or swabs.
3. Unwanted medicines and other pharmaceutical products.
4. Used syringes, needles and blades ("contaminated sharps").

Non-hazardous **domestic** medical waste (e.g. waste from small injuries or minor illnesses, soiled nappies, incontinence pads and sanitary towels), do not – under normal circumstances – constitute clinical waste, and may be disposed of as domestic rubbish.

Clinical waste is a category of hazardous waste, and has to be stored and collected under tightly controlled conditions and is generally disposed of by incineration. It cannot be put with normal rubbish. The process is legally enforced under the controlled waste regulations.

If your organisation generates clinical waste, ensure that the appropriate controls are in place to collect, store and dispose of it using licensed contractors.

Q

Do you have any clinical waste at your premises; how is it stored?

EXERCISE 3

Self-assessment questions

1. What is the primary symptom of *C. diff*?

....

....

....

2. What does MRSA stand for?

....

....

....

3. How long do the symptoms of Norovirus normally last?

a) 1–2 days
b) 1–2 weeks

4. List two symptoms of Tuberculosis.

a) ....

b) ....

5. Water is less likely to promote legionella bacteria growth if it is above which temperature?

a) 20°C
b) 25°C
c) 35°C
d) 45°C

6. List two of the three categories of falls from windows as detailed by the HSE.

a) ....

b) ....

7. Give two examples of sharps.

a) ....

b) ....

You will find the answers on pages 128–129.

MODULE FOUR

Accidents, incidents and infections

HOSPITAL AND CARE HOME BOSSES TO FACE PROSECUTION

DIRECTORS IN CHARGE OF CARE HOMES AND HOSPITALS CAN BE CHARGED WITH CRIMINAL OFFENCES IF THERE ARE FAILURES IN CARE AT THEIR ORGANISATIONS

Welcome to module four. In this module you learn more about accidents and incidents, how they occur and what needs to be done when they do occur.

WHAT IS AN ACCIDENT/INCIDENT?

These are unplanned/uncontrolled events which have led to, or could lead to, damage or injury to individuals, property, plant or any other loss to an organisation.

Accidents and incidents occur as a result of unsafe acts and unsafe conditions which are affected by:

Organisational factors

These are factors relating to the organisation you work for, for example:

- the safety culture within the organisation;
- management commitment to safety.

Human factors

These are factors that may relate to you; for example, your attitude, the training you have had, any disabilities you may have, your health and/or fitness.

Occupational factors

These are factors relating to the work you do and the environment you do it in, for example:

- If you deliver products there will be a greater risk of muscular strain due to the manual handling involved.
- If you work in a kitchen, you are more at risk from knife injuries and burns from touching hot surfaces.
- High noise levels in a factory can damage your hearing.
- Poor ventilation within a laundry or kitchen can cause heat stress.
- Poor lighting within an office can cause headaches and eyestrain.
- Working in a dusty atmosphere such as a commercial bakery can cause breathing problems.

Once things start to go wrong (i.e. we have an unsafe act and/or unsafe condition), they act like dominos. A chain reaction takes place that cannot be stopped, resulting in accidents and injury.

Domino effect leading to incidents

Each year there are thousands of accidents reported to the enforcement authorities. Some of the most common categories include:

- manual handling disorders
- slips, trips and falls
- being struck by a moving object.

Unfortunately 95% of all accidents are caused by human error and are therefore preventable.

Q

Can you think of any accidents that have recently occurred in your workplace? What do you think really caused them?

Once an accident, dangerous occurrence or near miss has occurred it is vital that it is reported for the following reasons:

1. Legislation

The Reporting of Injuries, Diseases and Dangerous Occurrences Regulations (RIDDOR) require employers to report the following to the local enforcing authority within specified time limits:

- fatal accidents
- fractures (not in the hand or foot)
- amputations
- loss of sight
- any injury resulting in immediate hospitalisation for more than 24 hrs
- specified dangerous occurrences such as a collapsing scaffold
- explosions
- specified workplace diseases that occur as a result of work activity.

The most common reason for employers to report an accident is when the injured person is off work for more than seven days as a result of the accident.

You as the employee need to notify your employer of all accidents that occur to you so that he can in turn fulfil his legal obligations.

2. Claim/Claim defence

In order for an employee to make a claim from their employer or from the Dept of Work and Pensions, there must be documented evidence of the incident having occurred; the accident book is often used to ensure this process takes place.

The employer also needs to know that the incident occurred so that he has the opportunity to do something about it. When reporting an accident, as much information as possible must be recorded.

3. Hazard identification

By documenting and being aware of the incident, the employer is better able to ascertain the cause of the incident and to implement corrective action to prevent it happening again (this may include carrying out a risk assessment or amending an existing risk assessment).

4. Monitor trends

Having records of all accidents that have occurred in a place of work over a period of time allows your employer to:

- analyse the types of incidents that have occurred;
- identify the staff involved and any training needs;
- ascertain the frequency of the incidents;
- ascertain the seriousness of the incidents;
- compare results with other similar businesses;
- compare results with previous years.

Research (carried out by Bird) has shown that for every accident resulting in a major injury there were approximately 10 resulting in minor injuries, 30 with property damage and 600 near misses.

WHY INVESTIGATE ACCIDENTS/INCIDENTS?

By investing time in investigating incidents your employer can:

- identify the cause of the accident/incident;
- improve safety standards within the workplace;

- prevent any recurrence by implementing a corrective action plan;
- defend fraudulent claims (employees have up to three years to claim for compensation);
- minimise the risk of prosecution by external enforcing authorities.

WHO SUFFERS FROM AN ACCIDENT?

The cost of having an accident can be counted in many ways. For example:

- loss of productivity
- plant/equipment damage
- physical harm
- psychological damage
- absenteeism
- loss of earnings
- permanent disability.

The real sufferers from the accident are the injured persons and their dependants. Always:

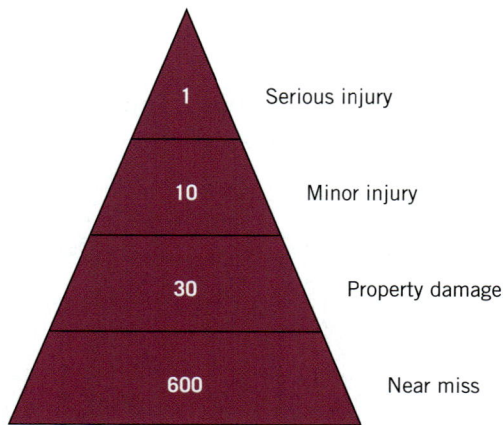

Bird's triangle

- **Think Safety** when carrying out your work
- **Work Safely** and you will be safe
- **Be Safe** and you will help avoid accidents involving yourself or others.

HAZARDS IN THE WORKPLACE

A definition of a hazard is:

"SOMETHING (E.G. AN OBJECT, A PROPERTY OF A SUBSTANCE, A PHENOMENON OR AN ACTIVITY) THAT CAN CAUSE ADVERSE EFFECTS."

In reality most things in the workplace or home have the potential to cause harm. For example:

A single brick placed on a flat table in a room is relatively safe; however, the same brick falling from some high scaffolding could hit someone below and cause serious injury or even death.

Damaged light fitting

A strong acid in a properly labelled closed glass container in a laboratory cupboard would be relatively safe; however, the same acid left in a cup in a kitchen could be accidentally drunk, causing serious burns to the mouth and tracts.

A simple change in situation or circumstances can change the nature of the hazard from safe to very unsafe. This is known as the risk associated with the hazard.

The definition of risk is:

"THE LIKELIHOOD OF THE HAZARD CAUSING HARM AND THE PROBABLE OUTCOME."

To help identification, hazards can be divided into five main categories within your workplace; these are:

1. **Physical:** These could be anything in the workplace you can feel or touch. Examples include:

   - doors
   - furniture
   - machinery
   - tools.

2. **Chemical:** These can be solid liquid or gas. Examples include:

   - detergents/soaps
   - acids/alkalies
   - correction fluid
   - bleach
   - glue
   - oils
   - paints
   - solvents.

3. **Biological:** This could be bacterial material on work surfaces. Examples include:

   - bacteria
   - moulds.

4. **Ergonomic:** Relates to the ill-health effects of poorly designed tasks and workstations leading to musculo-skeletal problems including work-related upper limb disorders, whole body or hand/arm vibration syndrome.

5. **Psychological:** This includes mental health, work-place stress, violence at work, smoking and drug/alcohol abuse.

Hazards can be managed or controlled in a number of ways. The following list of control measures (in order of preference) can be targeted at any hazard.

- Ideally we should aim to eliminate hazards all together by removing the hazard or changing the process.
- Substituting hazardous materials or processes for less hazardous ones.
- Keeping people away from the hazard or the hazard away from the people (i.e. preventing exposure).
- Limiting time spent in the hazardous area or carrying out the hazardous task helps limit the exposure to a hazard.
- Introducing local exhaust ventilation can help remove the hazard if it is gaseous in nature.
- Improving general ventilation in the workplace can help dilute the hazard if it is gaseous in nature.
- Improving housekeeping in the workplace can remove a number of tripping and falling hazards.
- Training employees to carry out the job correctly and safely being fully aware of the hazards will reduce the likelihood of accidents and injury.

- Ensuring employees have adequate welfare facilities can help to reduce the hazards.
- Where necessary and available, carrying out health screening can help ensure other control measures are working and detect problems early.
- Providing employees Personal Protective Equipment can help them reduce their exposure to the hazard providing that they are using the equipment in the right way.

Q

Could you, or does your employer, apply any of the above controls in your workplace?

Q

How would you describe your workplace? What types of hazard are you exposed to?

Keep all common areas clean and safe

OCCUPATIONAL HEALTH

A typical workplace today is very different to the workplaces present in the 1970s when the Health and Safety at Work Act was introduced. There has been a marked move away from heavy manufacturing industry and mining to light industry and service.

As a result, there has been a change in the types of injuries and diseases occurring. Although the legacy of some illnesses such as deafness, asbestosis, lung cancer that take years to manifest themselves are still present.

Although medical advances and technology have helped us reduce or even eliminate the risks associated with some hazards, it is unfortunate that we have in the last few decades introduced new hazards and risks. Of these, work-related stress appears to be one of the fastest growing illnesses.

A worker's health will be directly affected by the type of environment he is working in. Occupational health surveillance is an important tool that can be used by an employer as part of his management strategy to help monitor control measures in place and identify and diagnose potential health problems within the workforce early on so that they can be treated.

Examples of health surveillance include:

- audiometric tests for people working in noisy environments;
- blood lead levels for people working with lead;
- lung function tests for people working with asbestos;
- eye sight tests for VDU users.

Unfortunately because occupational ill health issues often take many years to manifest themselves, they are often ignored by employers and employees do not always realise they have a problem until it is too late.

EXERCISE 4

Self-assessment questions

1. What is the definition of risk?

....

....

....

2. Accidents and incidents occur as a result of what?

a) ....

b) ....

3. Give two examples of accidents that would be reportable to the enforcing authority.

a) ....

b) ....

4. List three different reasons for investigating accidents.

a) ....

b) ....

c) ....

5. What is the definition of a hazard?

....

....

....

6. List four of the five hazard categories.

a) ....

b) ....

c) ....

d) ....

7. Give two examples of health screening techniques.

a) ....

b) ....

You will find the answers on pages 129–130.

Proactive health and safety

TWO PARTNERS OF A CARE HOME FINED £50,000 EACH FOR THE DEATH OF AN ELDERLY RESIDENT

THE RESIDENT DIED FROM A HEAD INJURY AFTER BEING LIFTED FROM HER BED USING A HOIST. THE JUDGE SAID THERE HAD BEEN SIGNIFICANT FAILINGS IN THE MAINTENANCE OF THE HOIST AND SLING AND INSUFFICIENT TRAINING OF STAFF

Welcome to module five. In this module you will learn about some of the practical steps you and your employer can take to help improve health and safety within the workplace and therefore reduce the risk of accidents and ill health. Remember your employer can only succeed with your help, co-operation and commitment.

HEALTH AND SAFETY POLICY AND CLEAR RESPONSIBILITIES

Your employer must produce a written policy statement (if employing more than five staff) detailing his commitment to ensuring your and other people's health, safety and welfare. In addition he must set out clear responsibilities and accountabilities for key members of his staff to ensure that safety is really put into practice and make arrangements to ensure that all safety procedures developed are implemented, monitored and reviewed as necessary.

```
                              ┌─────────────────┐
            ┌─────────────────│     Policy      │
            │                 └─────────────────┘
            │                          │
            │                          ▼
            │                 ┌─────────────────┐
            ├─────────────────│   Organising    │
            │                 └─────────────────┘
            │                          │
            │                          ▼
   ┌─────────────┐           ┌─────────────────┐
   │  Auditing   │───────────│  Planning and   │
   └─────────────┘           │  implementing   │
            │                 └─────────────────┘
            │                          │
            │                          ▼
            │           ┌───────────────────────┐
            ├───────────│  Measuring performance │
            │           └───────────────────────┘
            │                          │
            │                          ▼
            │           ┌───────────────────────┐
            └───────────│  Reviewing performance │
                        └───────────────────────┘
```

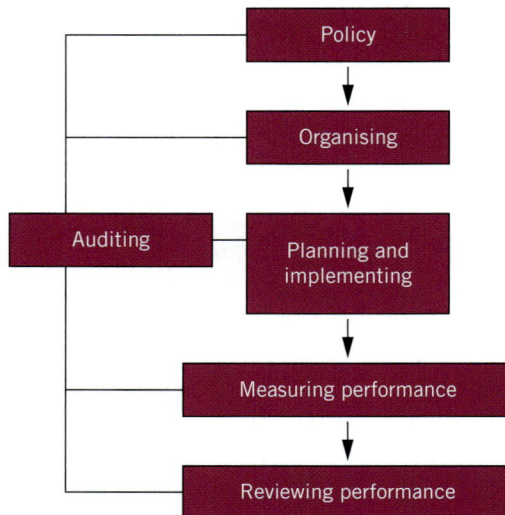

Successful health and safety management model (HSG 65)

AUDITS

Your employer may from time to time carry out audits to measure the effectiveness of the procedures and processes he has in place. These can help highlight deficiencies and gaps within his overall safety management system.

INSPECTIONS

Inspections may be carried out using checklists and/or proformas which measure conformance to the company's health and safety procedures in terms of documentation and the physical hazards present in the workplace. This allows your employer to "bench mark" and highlight the improvement or falling of standards ("performance tracking"). You may be asked to assist in the inspection process.

In reality you can carry out an inspection of your local place of work every day on an informal basis and by making your employer aware of any hazards, you could help to prevent an accident.

HAZARD/NEAR MISS REPORTING

As stated, your assistance is vital to improve safety standards; always report any hazards, near misses or dangerous situations you see or are aware of. In some instances you may even be able to resolve the issues yourself and help prevent an accident. Your employer may have a formal hazard/near miss reporting system in place; if so use it as it could help save a life or prevent an accident.

TRAINING

The purpose of training is to ensure that the employee is able to carry out the task/work activity to be performed consistently to an agreed standard and in a safe manner. All training, whatever the task, should cover any safety measures needed to ensure the task is performed safely (e.g. the correct use of PPE, manual handling techniques, machine operating).

Your employer may deliver the training on or off the job; it may be formal or informal, internal or external, in a group or individual basis. However, when training is conducted, the purpose is to ensure that you learn. Your employer should document details of any training provided to prove that it did take place.

Examples of when training is necessary include:

- induction training for all new employees to the business or new department/section
- refresher training
- specific training (e.g. first aid)
- training when new equipment is introduced.

An example of classroom training

SAFE SYSTEMS OF WORK/RISK ASSESSMENTS

A safe system of work is a formal written procedure which results from a systematic examination of a task or work activity in order to identify all the hazards and associated risks involved in the task or activity (risk assessment). Once completed it should be used to provide you with instruction in the safe method of work and to ensure that the hazards are eliminated or the remaining risks are minimised.

The risk assessment identifies the key steps within the activity (how it is done), all the potential hazards are then identified (the "what if" chain of events) and assessed for risk. This is then followed by an action plan of any preventative measures needed to eliminate or reduce the risks (the "what can be done to prevent accidents" part of the assessment).

Your employer must complete risk assessments for all tasks carried out, both routine and abnormal. Always ensure

The Process of Risk Assessment

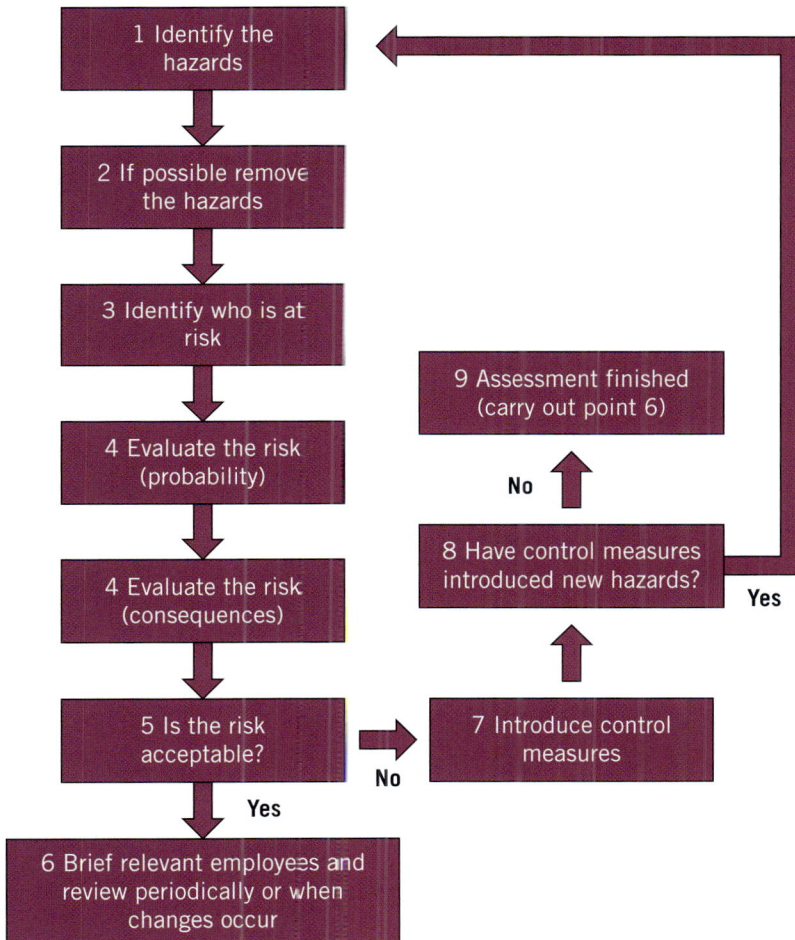

The risk assessment process flowchart

that you have been made aware of the relevant risks associated with the work you do and that you are aware of any particular precautions that must be taken.

PLANT/EQUIPMENT TESTING AND INSPECTION

Your employer has a duty to carry out or have carried out on their behalf specific tests and inspections to ensure that equipment/plant is safe to operate. Before using any plant or equipment, ensure you have the necessary training/ knowledge and always carry out a quick visual inspection of the plant or equipment prior to use.

COMMUNICATION

Your employer has a legal duty to consult with you on matters relating to health and safety, directly or through nominated representatives (these may be unionised workplace representatives or non-unionised workplace representatives). It is in his interest and yours to ensure that you are fully aware of all hazards and risks present within the workplace.

Remember communication is a two-way process. It may be achieved through many ways such as:

- formal meetings
- informal meetings
- newsletters
- tool box talks
- training sessions
- notice boards
- emails.

Ensure you know who your representatives are and how you can communicate matters relating to health and safety to your employer.

PLANNED MAINTENANCE

By developing and implementing a planned maintenance programme your employer cannot only comply with legislation but also help ensure that work equipment is likely to be safe to use when properly used.

CO-OPERATION

The management of health and safety is a two-way process requiring the commitment and co-operation of both employer and employee.

You can help your employer improve safety standards and reduce accidents by the positive actions you take.

PERSONAL HYGIENE

One of the most effective and simple ways of protecting yourself and helping to prevent the spread of infection is the process of thorough and regular hand washing and the avoidance of touching parts of your own body unnecessarily (e.g. mouth, nose).

Unfortunately many people fail to wash their hands as often as they should. Common reasons for not washing hands include:

- skin irritation/drying out of the skin
- hand washing facilities not readily available
- already wearing gloves
- too busy
- lack of appropriate staff around to assist
- lack of awareness of need/benefits
- being a physician.

Hands should be washed:

- if your hands look dirty;
- before and after any activity that could have dirtied

your hands, even if they look clean, such as after going to the toilet and before and after preparing food;
- before and after every activity or procedure involving contact with a patient and between patients, before contact with your body (if you are the patient), or contact with equipment;
- if you are caring for more than one person, wash your hands in between looking after each person;
- even if you propose to wear gloves.

How to wash your hands:

- Cover any cuts or grazes with a waterproof plaster.
- Keep your fingernails short, clean and free from nail polish.
- Take off your watch and any jewellery such as bracelets or rings (if you can).
- Wet your hands under luke-warm running water.
- Use a liquid soap or antimicrobial (anti-germ) solution and water.
- Make sure the handwash you are using covers all the surfaces of your hands.
- Rub your hands together vigorously for at least 10 to 15 seconds, remembering the tips of your fingers, your thumbs and the areas between the fingers.
- Rinse your hands with warm water and dry them with good quality paper towels that are absorbent and soft.

If you are using an alcohol-based handrub then:

- make sure the solution used covers all the surfaces of your hands. Rub them together vigorously, remembering the tips of your fingers, your thumbs and the areas between the fingers;
- rub it in until it has evaporated and your hands are dry;

Always wash your hands thoroughly

- remember alcohol-based handrubs do not provide protection against some viruses.

Using a suitable moisturising hand cream regularly can help to prevent dry hands and damage to skin.

Always tell your manager if a particular soap or cleaning product irritates your skin.

PERSONAL PROTECTIVE EQUIPMENT

Sometimes you may need to use personal protective equipment to stop infection passing between you and the person you are caring for or something you are handling. In these situations always ensure that

- when using gloves which act as an additional barrier you should still wash your hands before and after you use them;

- if you are sensitive to rubber, or experience a skin reaction using gloves, tell your manager;
- your gloves are put on before having any contact with the inside of the body (including inside the mouth) or with a wound, or if you are carrying out an activity that might lead to contact with blood or body fluids or with sharp or dirty instruments;
- each pair of gloves should only be used once;
- gloves should be changed between patients, and between different activities or procedures for the same patient;
- gloves are safely disposed of after use;
- when using disposable plastic aprons (where there is a risk of body fluids or blood splashing onto your clothes), wear the apron for one procedure only;
- the apron is safely disposed of after use;
- after disposal of the apron always wash your hands.

From time to time it may be necessary to use face masks and eye protection where there is a risk of fluids splashing into your face or eyes.

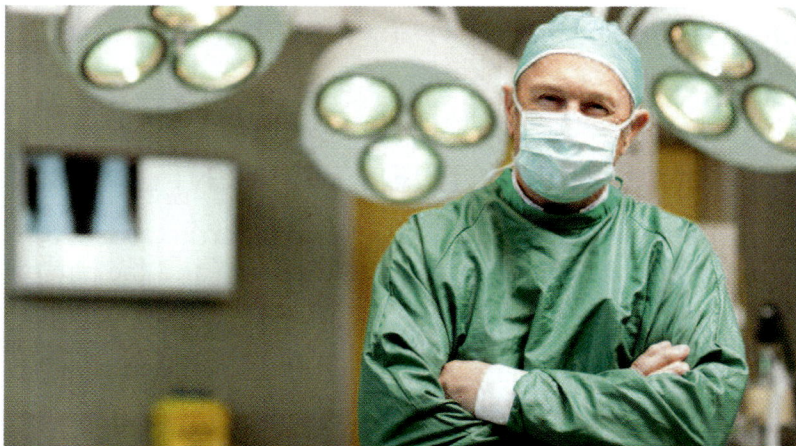

Always wear the correct PPE

EXERCISE 5

Summary questions

To assess your level of understanding please complete the following exercise.

1. Give two examples of how inspections can help manage health and safety.

   a) ....

   b) ....

2. Is there a legal requirement to carry out risk assessments?

   a) Yes
   b) No

3. List three different ways of delivering training.

   a) ....

   b) ....

   c) ....

4. List three ways the employer can communicate with employ-ees.

   a) ....

   b) ....

   c) ....

5. Why is it important to report hazards and near misses?

....

....

....

6. Does your employer have to produce a safety policy statement?

a) Yes
b) No

7. When is it necessary to provide training?

....

....

....

You will find the answers on pages 130–131.

Answers to questions

MODULE 1 – Health and safety law and enforcement

1. What is the maximum fine you can receive at a magistrates' court?

c) £20,000

2. What are the two important types of law used by the CQC?

c) Civil & Criminal

3. Can an enforcement officer take equipment away from your workplace?

a) Yes

4. List four reasons for employers to address health and safety.

Moral, Legal/Law, Financial/Cost, Business need

5. Give two reasons why employees should take health and safety seriously.

Legal requirement, moral, financial benefits, business needs

6. Who enforces health and safety law in the UK?

c) Environmental Health Officers and Factory Inspectors

7. Health and safety law places a duty on:

c) Employers, employees and the self-employed

MODULE 2 – Health and safety regulations

1. What are the four colours used for safety signs?

Blue, Yellow, Red and Green

2. Is it good practice to lift an adult person who you find fallen to the floor on your own?

b) No

3. What are the two types of radiation?

Ionising and non-ionising

4. List four different health and safety control measures that can be applied to reduce risk.

Elimination, substitution, limit exposure, local exhaust ventilation, health surveillance, general ventilation, housekeeping

5. Which muscles should be used when lifting a load?

b) Leg muscles

6. What is the purpose of First Aid?

To preserve life and prevent the deterioration of the casualty

7. List two things that must be assessed when carrying out a VDU assessment.

The screen, keyboard, chair, desk, environment

MODULE 3 – Commonly acquited infections and common hazards

1. What is the primary symptom of *C. diff*?

Diarrhoea

2. What does MRSA stand for?

Methicillin-resistant *Staphylococcus aureus*

3. How long do the symptoms of Norovirus normally last?

a) 1–2 days

4. List two symptoms of Tuberculosis:

Persistent cough for more than three weeks that brings up phlegm, which may be bloody, weight loss, night sweats, high temperature (fever), tiredness and fatigue, loss of appetite

5. Water is less likely to promote legionella bacteria growth if it is above which temperature?

d) 45°C

6. List two of the three categories of falls from windows as detailed by the HSE.

Accidental falls, falls arising out of a confused mental state, deliberate self-harm

7. Give two examples of sharps:

Needles, lancets, scalpels, stitch cutters, glass ampoules, sharp instruments, broken crockery and glass

MODULE 4 – Accidents, incidents and infections

1. What is the definition of risk?

The likelihood of a hazard causing harm and the probable consequences

2. Accidents and incidents occur as a result of what?

Unsafe acts, unsafe conditions

3. Give two examples of accidents that would be reportable to the enforcing authority.

Fractures, amputations, death, over three-day absence injury, 24-hour hospitalisation, unconsciousness, loss of sight

4. List three different reasons for investigating accidents.

Identify cause, prevent recurrence, claim defence, insurer stipulation

5. What is the definition of a hazard?

Something that can cause adverse effects

6. List four of the five hazard categories:

Physical, chemical, biological and ergonomic or psychological

7. Give two examples of health screening techniques.

Audiometry, lung function testing, chest x-rays, blood sampling

MODULE 5 – Proactive health and safety

1. Give two examples of how inspections can help manage health and safety.

Identify hazards, prevent accidents, demonstrate commitment

2. Is there a legal requirement to carry out risk assessments?

Yes

3. List three different ways of delivering training.

One to one, on the job, off the job, group training, distance learning, computer-based training, verbal instructions

4. List three ways the employer can communicate with his employees.

Tool box talks, notice boards, newsletters, verbal instructions, meetings, videos

5. Why is it important to report hazards and near misses?

To prevent them from beccming accidents in the future

6. Does your employer have to produce a safety policy statement?

Yes

7. When is it necessary to prcvide training?

Induction, change of job, promotion, refresher, new equipment or change in process

Glossary/Definitions

Accident An unplanned and/or uncontrolled event which has led to injury, damage or other loss to the business

COSHH Control of Substances Hazardous to Health Regulations

dB Decibel, the measure of sound

DSE Health and Safety (Display Screen Equipment) Regulations

EHO Environmental Health Officer

EMAS Employment Medical Advisory committee

EPA Environmental Protection Act

FFI Fee for intervention

Hazard Something (e.g. an object, a property of a substance, a phenomenon or an activity) that can cause adverse effects

H&S Health and safety

HSC Health and Safety Commission

HSE Health and Safety Executive

HSWA Health and Safety at Work Act

MHOR Manual Handling Operations Regulations

MHSWR Management of Health and Safety at Work Regulations

MSDS Material safety data sheet

NICE National Institute for Health and Care Excellence

PAT Portable appliance testing

PPE Personal protective equipment

PUWER Provision and Use of Work Equipment Regulations

RIDDOR Reporting of Injuries, Diseases and Dangerous Occurrences Regulations

RISK The probability and consequences of the hazard occurring

RPE Respiratory protective equipment

SHE Safety, Health and Environment

Six Pack Term given to a group of six regulations introduced in 1993: **MHSWR, MHOR, PPE, DSE, PUWER, WHSWR**

UK United Kingdom

VDU Visual display unit

WEL Workplace exposure limit

WHSWR The Workplace (Health, Safety and Welfare) Regulations

WRULDs Work-related upper limb disorders

Image credits

Figure 1 Injured man © Rtimages

Figure 2 Winning trophy © Shutterstock bioraven

Figure 3 Employers and employees must work together © Tsyhun

Figure 4 Reasonably practicable © Subash Ludhra

Figure 5 Failing to manage your health and safety can affect your employee's, patients' or client's well-being © jvinasd Shutterstock

Figure 6 Obey the law and your duties © Shutterstock my portfolio

Figure 7 County court © Subash Ludhra

Figure 8 Magistrates' court © Shutterstock Stuart Monk

Figure 9a An example of a mandatory requirement sign © Subash Ludhra

Figure 9b An example of a prohibition sign © Subash Ludhra

Figure 9c An example of a warning sign © Subash Ludhra

Figure 9d An example of a safe condition sign © Subash Ludhra

Figure 10 Do not overload sockets/extension leads © Shutterstock Kordik

Figure 11 Cleaning chemicals © Shutterstock mertcan

Figure 12 Typical first-aid sign © Shutterstock Atlaspix

Figure 13 Ceiling hoist © Subash Ludhra

Figure 14 Mobile lifting hoist © Shutterstock daseaford

Figure 15 Always wear the correct PPE © Shutterstock Diana Valujeva

Figure 16 Hospital ward © Shutterstock wavebreakmedia

Figure 17 Clear all spillages as quickly as possible and erect warning signs as necessary © Shutterstock hxdbzxy

Figure 18 Bedside guarding © Shutterstock Morphart Creation

Figure 19 Display screen equipment workstation © Shutterstock trekandshoot

Figure 20 X-ray machine © Shutterstock Poznyakov

Figure 21 Keep air conditioning towers maintained © Shutterstock Wichan Kongchan

Figure 22 Follow the correct procedures when fitting or removing catheters © Shutterstock Sherry Yates Young

Figure 23 Follow the correct procedures when fitting or removing catheters © Shutterstock Henk Vrieselaar

Figure 24 Ensure patients cannot fall from windows © Shutterstock bikeriderlondon

Figure 25 Sharps box © Shutterstock Jamie Rogers

Figure 26 Domino effect leading to incidents © Shutterstock Ljupco Smokovski

Figure 27 Bird's triangle

Figure 28 Damaged light fitting © Subash Ludhra

Figure 29 Keep all common areas clean and safe Shutterstock Rob Marmion

Figure 30 Keep all work areas clean and safe © Shutterstock Lisa S.

Figure 31 Successful health and safety management model (HSG 65) © Subash Ludhra

Figure 32 An example of classroom training © Subash Ludhra

Figure 33 The risk assessment flowchart © Subash Ludhra

Figure 34 Always wash your hands thoroughly © Shutterstock bikeriderlondon

Figure 35 Always wear the correct PPE © Shutterstock wavebreakmedia

Examination

Once you have completed the guide you can choose to complete the examination.

Complete the following exam to test your international health and safety workplace knowledge. All candidates will be informed whether they have passed or failed via email, and candidates achieving a pass mark of 75% or above will receive a certificate.

To find out details of where to send your completed exam, or to request an electronic version, visit www.routledge.com/9780415835466.

You have thirty minutes – good luck.

Your FULL Name: (PRINT)		Employee Reference:	Employer Name:
Date of Birth:		Your Email Address:	
Date of Examination:			

Please indicate by putting a tick in the appropriate box(es) your choice of correct answer(s)

1. What are the **TWO** important types of law used by the CQC?

    a) Criminal ☐
    b) Medical ☐
    c) Civil ☐
    d) Advisory ☐

2. Under the HASAW Act, what are the **TWO** enforcement notices that are issued

    a) Confiscation order ☐
    b) Prohibition ☐
    c) Renewal ☐
    d) Improvement ☐

3. What is the maximum fine that can be imposed for a health and safety offence in a **magistrates' court**?

    a) £5,000 ☐
    b) £10,000 ☐
    c) £20,000 ☐
    d) £40,000 ☐

4. What is meant by the term **reasonably practicable**?

    a) The balance between risk and health ☐
    b) The balance between risk and cost ☐
    c) The balance between cost and health ☐
    d) The balance between cost and prosecution ☐

5. Which one of the following **is not** a duty of an employer under the Health and Safety at Work Act?

    a) Providing information, instruction and training for employees ☐
    b) Providing a safe place of work ☐
    c) Providing safe systems of work ☐
    d) Consulting with employees ☐

6. Which one of the following **is not** a duty of an employee under the Health and Safety at Work Act?

    a) To buy personal protective equipment for themselves ☐
    b) Not to interfere or misuse anything provided for health and safety ☐
    c) Care for the health and safety of themselves and other people ☐
    d) Co-operate with his employer ☐

7. What does MRSA stand for?

   a) Methicillin-refined *Staphylococcus aureus* ❒
   b) Methicillin-resistant *Staphylococcus aureus* ❒
   c) Methicillin-resistant *Staphylococcus aroma* ❒
   d) Measles-resistant *Staphylococcus aureus* ❒

8. Water is less likely to promote legionella bacteria growth if it is above which temperature?

   a) 17°C ❒
   b) 27°C ❒
   c) 37°C ❒
   d) 45°C ❒

9. What **message** does a Red safety sign give?

   a) Warning ❒
   b) Mandatory ❒
   c) Prohibition ❒
   d) Safe condition ❒

10. Which of the following are classed as sharps?

   a) Needles ❒
   b) Broken glass ❒
   c) Scalpels ❒
   d) Blunt scissors ❒

11. Which **muscles** should be used when lifting a load?

   a) The arm muscles ❒
   b) The leg muscles ❒
   c) The back muscles ❒
   d) The stomach muscles ❒

12. What are the **main purposes** of first aid?

    a) To cure the casualties, illnesses ☐
    b) To preserve life and prevent the deterioration of the
       casualty ☐
    c) To injure the casualty ☐
    d) To perform miracles ☐

13. Which **four factors** must be taken into account under the manual handling operations regulations?

    a) The task, load, working environment and individual
       capability ☐
    b) The weight, size, temperature and dimensions ☐
    c) The temperature, room space, individual capability
       and colour ☐
    d) The task, load, shape and time ☐

14. What are the **dangers** associated with electricity?

    a) Death ☐
    b) Electrical burns and shock ☐
    c) Fires ☐
    d) All of the above ☐

15. What is the **most effective** way of controlling a hazard and its associated risk?

    a) Prevent exposure to it ☐
    b) Substitute it ☐
    c) Eliminate it ☐
    d) Provide personal protective equipment ☐

16. What is the **minimum** acceptable temperature in an office?

    a) 25°C ☐
    b) 11°C ☐
    c) 16°C ☐
    d) 21°C ☐

17. Which **one** of the following injuries is **not** reportable to the enforcing authorities?

    a) Death ☐
    b) Fractured arm ☐
    c) Amputated leg ☐
    d) Fractured finger ☐

18. Which **one** of the following items **must** be assessed when carrying out a display screen equipment assessment?

    a) The operator's clothing ☐
    b) The keyboard ☐
    c) The operator's shoes size ☐
    d) The number of plants in the room ☐

19. What is the **definition of risk**?

    a) Something that will cause harm ☐
    b) Something that has the potential to cause harm ☐
    c) The probability of harm being caused and the likely consequences ☐
    d) Being struck by a falling brick ☐

20. What is the **definition** of a **hazard**?

   a) Something that will cause harm ☐
   b) Something that can cause adverse effects ☐
   c) The probability of harm being caused and the likely
      consequences ☐
   d) Being struck by a falling brick ☐

21. Which **one** of the following is **not** a typical example of occupational health monitoring?

   a) Audiometric testing ☐
   b) Lung function testing ☐
   c) Blood lead level testing ☐
   d) Employee weight checks ☐

22. Which **one** of the following is **not** a primary reason for investigating accidents?

   a) Defending a claim for compensation ☐
   b) Identifying the cause of the accident ☐
   c) Helping to prevent a reoccurrence of the accident ☐
   d) Apportioning blame on employees ☐

23. **How long** does an employee have to make a compensation claim following an accident at work?

   a) One year ☐
   b) Three years ☐
   c) Five years ☐
   d) Ten years ☐

24. What is clinical waste?

    a) Human or animal tissue   ❐

    b) Dressings and swabs   ❐

    c) Cardboard waste   ❐

    d) Unwanted medicines   ❐

25. **How long** do the symptoms of Norovirus normally last?

    a) 1–2 days   ❐

    b) 3–4 days   ❐

    c) Less than one day   ❐

    d) More than 5 days   ❐

26. Which one of the following is **not** a proactive safety measure?

    a) Carrying out regular audits   ❐

    b) Carrying out regular training   ❐

    c) Carrying out regular workplace inspections   ❐

    d) Investigating an accident   ❐

27. When is it **not** appropriate for an employer to provide health and safety training for his staff?

    a) When starting a new job   ❐

    b) After an accident   ❐

    c) When a new machine has been introduced to the workplace   ❐

    d) When leaving the company   ❐

28. Which of the following safety measures is an example of **reactive monitoring**?

a) Carrying out regular audits ☐
b) Carrying out regular training ☐
c) Carrying out regular workplace inspections ☐
d) The analysis of accident data ☐

29. According to Bird's triangle **how many** near misses equate to one serious injury?

a) 500 ☐
b) 600 ☐
c) 450 ☐
d) 700 ☐

30. Which of the following **are** typical symptoms of Tuberculosis?

a) Weight loss ☐
b) High temperature (fever) ☐
c) Night sweats ☐
d) Diarrhoea ☐

Well done. You have now finished the test. Please check to ensure that you have answered all 30 questions (remember some questions require two answers).